Earth, Fire and Water: Chinese Ceramic Technology

A Handbook for Non-specialists

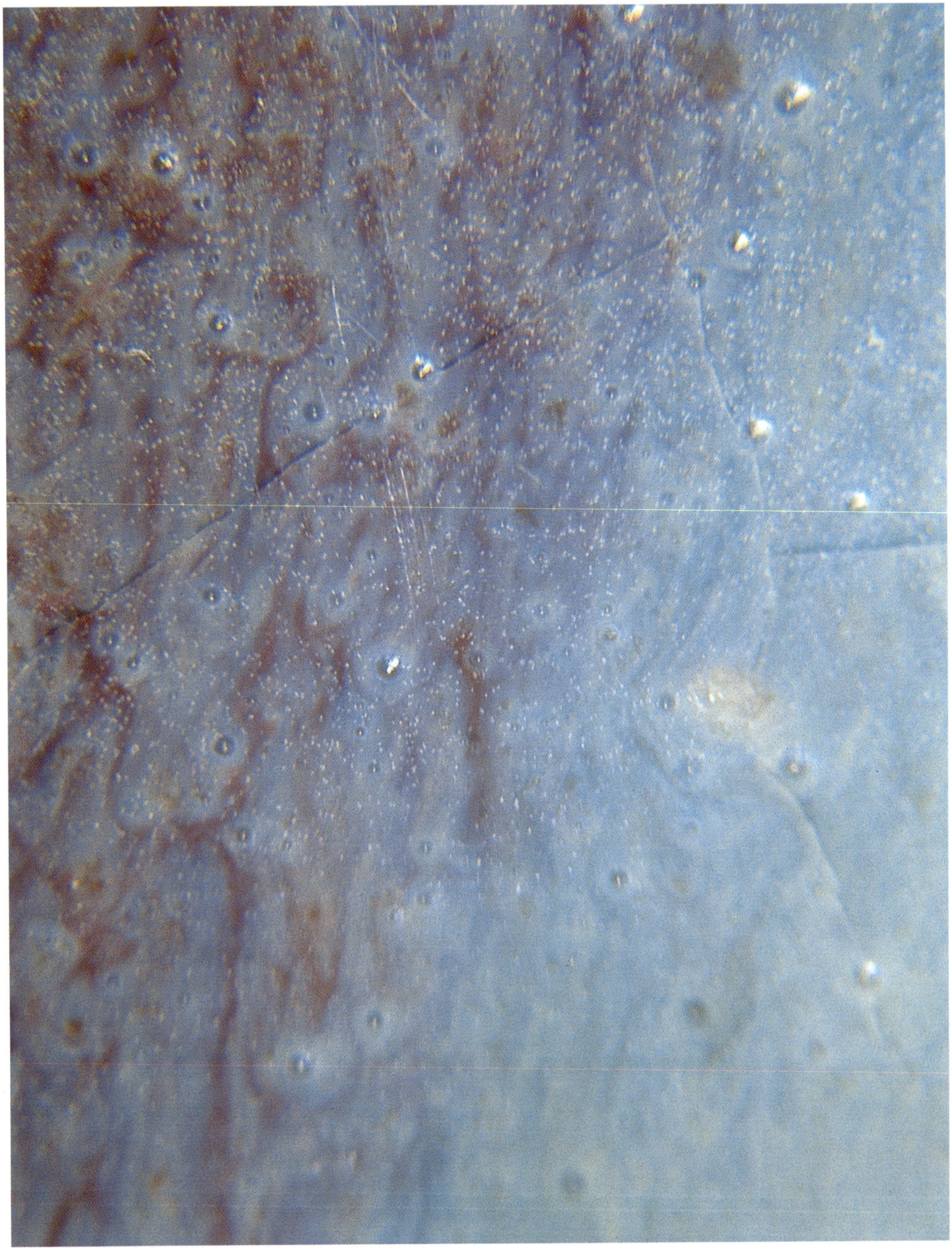

Detail of a Jun ware glaze

Earth, Fire and Water: Chinese Ceramic Technology

A Handbook for Non-specialists

by
Stacey Pierson

UNIVERSITY OF LONDON
PERCIVAL DAVID FOUNDATION OF CHINESE ART
School of Oriental and African Studies
53 Gordon Square
London WC1H 0PD

Front cover:

92	VASE, JUN WARE Northern Song dynasty, early 12th century
A4	LOBED DISH, JUN WARE Northern Song dynasty, 12th century

Back cover:

320	JAR WITH UNDERGLAZE IRON DECORATION Tang dynasty, dated by inscription to AD 823
A696	EWER WITH UNDERGLAZE RED DECORATION Ming dynasty, Hongwu period (1368-98)
674	FLASK WITH UNDERGLAZE BLUE DECORATION Ming dynasty, early 15th century
A678	BOWL WITH UNDERGLAZE BLUE AND RED DECORATION Ming dynasty, Xuande mark and period (1426-35)

Published by the School of Oriental and African Studies, University of London

Printed By: Balding and Mansell
ISBN 0 7286 0265 2

Photography by Glenn Ratcliffe

Contents

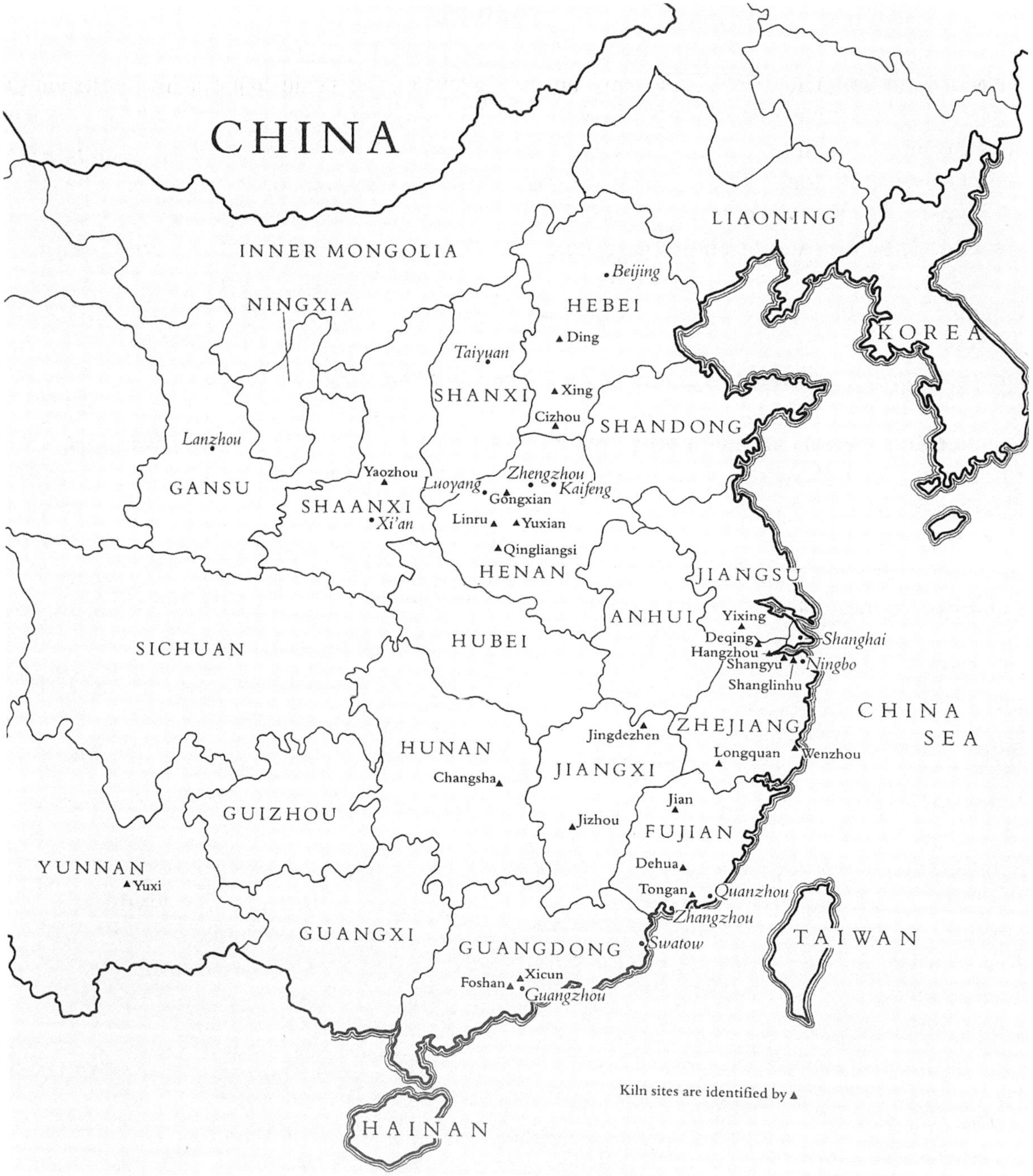

Map of China with kiln sites

Preface

This book was written to accompany an exhibition of the same name at the Percival David Foundation of Chinese Art in 1996. Both the exhibition and this handbook were inspired by the many requests from students and *Friends* of the museum for basic information about ceramic technology that does not require experience of making pots or a degree in chemistry to understand. In the past decade, the study of art history has changed quite dramatically in that technology is now an important part of the formal study of works of art. Knowing how a work of art is made and what materials were used in its manufacture can reveal much about its history and its visual appearance. This is especially true of Chinese ceramics.

In Britain and in China, the study of Chinese ceramics often falls into the category of archaeology, not art history. The reasons for this are too numerous to discuss here but in practice, this means that much of the material published on Chinese ceramics is in fact scientific in nature. It is not uncommon for exhibition catalogues to include quite technical articles by leading ceramic scientists, archaeologists or potters, which often discuss new information or discoveries in language that is unintelligible to the average student or collector. The aim of this book is therefore to explain the basic science involved in ceramic manufacture with reference to Chinese ceramics, as they are among the most technologically advanced and thus influential ceramics in the world. This is not an exhibition catalogue but rather a basic reference work for collectors, art historians and anyone who is interested in Chinese ceramics. It assumes no previous knowledge of ceramic technology or science in general and will use only terms that can be defined and explained in a manner accessible to the non-specialist. This will, I hope, enable the reader to understand most of what is written about Chinese ceramics and will illustrate the impact of technology on the visual appearance of the main types of ceramics produced in China.

Throughout the book the raw materials used and methods of manufacture employed in the making of a range of ceramics will be discussed and then put into context, giving a history of ceramic development in China. For simplicity, the text is divided into five chapters, each focusing on an aspect of ceramic technology: Clays and Their Origins, Glazes, Decorative Techniques and Forming Methods, Kilns and Kiln Furniture and Drying and Firing. Much of what is discussed in this book is applicable to ceramics in general. However the chapter on decorative techniques concentrates on the main types used on Chinese ceramics and so may not be applicable to other ceramic traditions. For reasons of space, not every type of ceramic made in China will be discussed but the most important wares of each main period will be included. For those who wish to explore the subject further, a selected bibliography has been provided which includes more detailed works on ceramic science and technology as well as references to specific wares or glazes.

Note: As this book has been designed for the general reader with little or no background in science or practical pottery, certain principles have been simplified in the interests of understanding and thus may not be strictly correct. A number of points have been summarized but can be explored in detail in several of the books in the bibliography.

Stacey Pierson
Assistant Curator

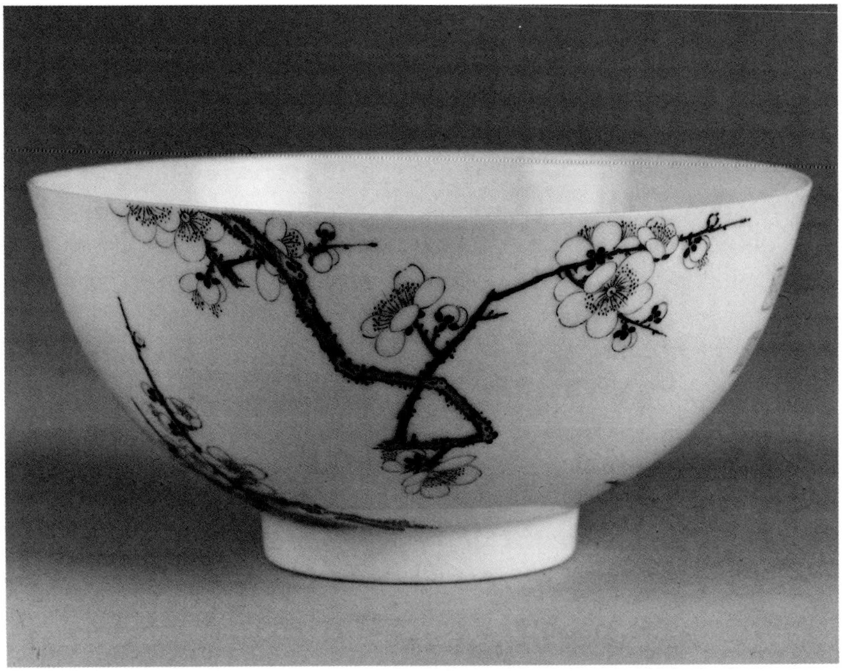

1. Bowl with overglaze enamel decoration in *Guyue xuan* style, Qing dynasty, Yongzheng mark and period (1723-35). PDF 827.

1. Clays and Their Origins

Ceramics and objects made from clay are part of our daily lives. From coffee cups and lamps, to bricks and waste pipes, ceramic materials have become a necessary part of everyday living. Because ceramic objects are so familiar, however, we tend not to think about how they are made and what they are made from, i.e. clay. A Ming vase, for example, is the end result of a long series of processes: mining clay, preparing it, forming the clay, decorating it, glazing and then baking in a kiln. As clay products, ceramics, therefore, are a natural corollary of the Earth's geology. Clay is found all over the world and comes in many different types and colours. It is the fundamental raw material used to make all ceramic wares.

Most of us have played with clay at sometime in our lives but probably have never really thought about what it actually is. We know that it is a soft, moist, earthy substance that can be formed into different shapes but it is certainly different from mud or wet sand. In fact, clay is a kind of rock. It is a soft, sedimentary rock that is formed from the breakdown of hard, igneous or "fire rocks" like granite or basalt. As a rock, clay is composed of minerals, in particular a group of minerals called "clay minerals", the most common of which is kaolinite. These minerals belong to a group called alumino-silicate minerals because they are composed of alumina and silica. These two elements are therefore the main components of any ceramic. (Note: as a general rule, substances with the suffix "ite" are classified as minerals.)

Clay and other rocks are formed when granite or basalt, the two rocks that make up the Earth's crust or surface, is weathered by wind, rain or hydrothermal reactions (underground water) and then decomposes or rots and forms deposits. Granite, the most plentiful igneous rock, is composed of quartz, mica and feldspar, all silicates or minerals based on silica.[1](figure 4) During decomposition, these components are liberated so that bits of quartz, mica or feldspar rocks can then be found deposited elsewhere. In the case of feldspar, if it too is subjected to weathering, it decomposes into kaolinite, the most common clay mineral. Feldspar itself is composed of alumina, silica and either soda or potassia so that when it decomposes, the extra ingredients are dispersed and we are left with pure alumina and silica, which together form kaolinite.[2](figure 6). With the addition of water, kaolinite and the other clay minerals form clay.

---

[1]Granite is formed when magma or molten rock from the earth's core cools and crystallizes then pushes through the earth's surface. It is therefore commonly known as a "trapped" or "intrusive" rock.

[2]Kaolinite also contains water in the form of H_2O. Its chemical formula is: $Al_2O \bullet 2SiO_2 \bullet 2H_2O$.

It is the soda or salt and excess silica that disperse and settle during the breakdown of granite and feldspar which become the salt in the ocean and sand at the beach.

Because of its geological origins, clay is often classified in the same way rocks are. One of these classifications is known as the "depositional situation" or where it was deposited during weathering. If you imagine that wind and rain as well as streams can carry rocks quite a distance, then where they deposit these rocks becomes important. If a rock or clay is deposited close to its "parent" rock, in this case the granite, then it is called a primary or residual clay. Its composition is therefore very close to that of the parent rock. If it is carried a long way or transported and deposited far from the parent rock, then it is known as a secondary or sedimentary clay. Sedimentary deposits contain materials from other sources besides the parent rock so the composition of sedimentary clays is quite different from that of the parent rock. This carrying action does have an impact on the type of clay produced. For example, if a sample of rock debris has been carried by stream for a very long way it has been subjected to quite a lot of water pressure and grinding from other rocks in the stream. The clays formed from it have particles that are finer than those of primary clay. They will also have fewer impurities.[3] The formation of clays can been seen as part of the continual process of transformation of the rocks on the earth's crust.

Another more familiar way of classifying clays is by practical potters' definitions which are based on the type of ceramic body they produce: earthenware, stoneware and porcelain. (Note: The ceramic "body" is the part that is made of clay, as opposed to the glaze or surface decoration.) Earthenware bodies are low fired (to approximately 1000°C) and are therefore rather soft and porous. This is why they make such good flower pots because they can absorb water. They can be red firing as in terracotta or white firing. Red earthenwares contain a lot of iron and have probably been formed from the decomposition of basalt, a black igneous rock that contains significant amounts of iron.(figure 7) Stoneware bodies are hard and non-porous and are fired to approximately 1200°C. They can be any colour but are typically grey or pale cream or white, depending on how much iron is present in the clay.(figure 8) Porcelain bodies are very hard, white and translucent. They are fired to the highest temperature, up to 1350°C and are very pure.(figure 9) It should be noted that these potters' classifications are particular to the West. In China, they usually do not apply. Chinese ceramic historians and archaeologists traditionally recognize only two types of bodies: high-fired (*ci*) and low-fired (*tao*). Thus in China, there is often no distinction between stoneware and porcelain. This sometimes leads to confusion when a celadon, or ceramic with a green glaze, such as Longquan ware for example, is described as a porcelain when translated from Chinese, but is called a stoneware in English. The Longquan

---

[3]See chapters two and three in the book *Ceramic Technology for Potters and Sculptors* by Cuff for a full explanation of clay formation. This book is listed in the bibliography.

body is made from a substance called "china stone" (see discussion below), a key ingredient in porcelain, but is not porcelain because it is not translucent and usually contains significant amounts of iron which ensure that the body is not pure white.

Another classification of clays and ceramic wares is particular to China and refers to the geographical origin of a clay or ceramic ware. When discussing Chinese ceramics, it is important to know whether a certain ware is "northern" or "southern". This rather arbitrary division roughly refers to those wares made north of the Yangtze river and those made south of the river. This river roughly divides China in half geographically and geologically, the clays in the north being quite different from those in the south. In general, Chinese ceramics are made from two main ingredients: "china clay" or kaolin and "china stone" or petuntse (*baidunzi*). The chemistry of these two materials will be discussed below but with regards to the geographical origins of Chinese ceramics, those made in the north tend to have bodies composed of mainly china clay or other clays. They are "clay rich" bodies. Those made in the south, however, have bodies made mainly of china stone. Clay rich bodies tend to be dense and compact whereas southern china stone based bodies tend to be quite glassy and sugary in texture. In practice, ceramics in the north are higher firing than those in the south because they contain more clay and thus more alumina which melts at a higher temperature than silica. The southern body is very high in silica and low in alumina which is why it has a sugary, glassy texture. Silica, of course, is the main ingredient in glass.

As we have seen, ceramics are made from clay and not mud or sand and thus clay has particular properties which enable it to be transformed into ceramics which mud and sand do not possess, namely, the ability to be formed into a shape and to hold that shape when fired. One of the most fascinating aspects of ceramics is the fact that they are made by one substance being transformed into another. In this respect, ceramics are very much like baking and baking analogies can be quite useful in explaining ceramics and ceramic processes. Like baking, dry ingredients, water and heat are needed to make ceramics, hence the title of the PDF exhibition, Earth, Fire and Water. In baking, it is usually eggs that act as the catalyst for transformation. In ceramics, however, it is the particular structure of the clay mineral. As noted above, clay minerals are crystalline in structure. These crystals look like little plates stacked on top of one another and stacks of crystals or "platelets" are stuck at right angles to other stacks. In a mass of kaolin, or white clay composed mainly of kaolinite, these stacks of crystals are pushed around when a shape is formed from the clay and thus are aligned in layers which can slide over one another with the addition of water, a characteristic known as "plasticity". By adding water to what is essentially dry crystals, one can make a substance that can be formed in to a shape and will hold that shape, much like American cookie dough which is made by adding liquid to dry ingredients like flour and sugar. In ceramics, this added liquid is water and is called the "water of plasticity". By heating the clay and thus gradually removing the added water, one can force the stacks of clay minerals to move closer and closer together and eventually to fuse, making a hard substance or ceramic.

Plasticity is a term that is frequently used in describing ceramics and it is often misunderstood. It is primarily a function of particle size in clay. The smaller the particles, the more water can move between them and the more easily the particles can slide across one another. It is similar to workability but slightly different in that a plastic clay may be easier to form on the wheel for example because the clay particles move easily over one another but the clay can be slightly floppy and thus not particularly "workable". A non-plastic clay will be harder to form into a shape but it will hold that shape better. A useful analogy is again taken from baking. A plastic clay is very much like bread dough, it can be stretched to several times its original length without breaking. A non-plastic clay is more like short-crust pastry in that it stretches slightly and then breaks quite easily and crumbles. When forming shapes from bread dough, one can make a wide range of shapes but not particularly tall or elaborately shaped ones as they will quickly collapse. The opposite is true of short-crust pastry which can be pinched, carved and moulded and will stay that way without much support. Accordingly, non-plastic clays are often described as "short" clays. The ideal clay body, therefore, would be one that is a mixture of plastic and non-plastic clays so that it can be formed quite easily into a shape and then will hold that shape during drying and firing.

As noted above, Chinese ceramics are generally made from two main ingredients: China clay and china stone. China clay or kaolin gets its name from *Gaoling*, a mine site in south China near Jingdezhen where there was a large deposit of china clay. Kaolin is a white firing clay that is primarily composed of kaolinite and is not as plastic as other clays because kaolinite has the largest particles of all the clay minerals. Another body material that one finds at least in early Chinese ceramics is loess, which is a kind of soft rock dust found in north China which can contain some clay minerals. It is composed mainly of quartz, mica, feldspar and calcite or limestone and could be used to produce low-fired earthenware bodies. It was used in neolithic ceramics, the "terracotta army" and in Han and some Tang earthenwares. Loess was also used for making cast bronze moulds which relied on its low shrinkage and fine texture to produce elaborately decorated bronze vessels. Loess is perhaps most familiar as the building material "adobe" or brickearth.

The structure of china clay (kaolin), has already been discussed with respect to kaolinite but that of china stone has not and it is useful to compare the two. As we have seen, southern Chinese ceramic bodies are made mainly from china stone. As its name suggests, china stone is a type of rock, like china clay, and it is composed of quartz and sericite or potassium mica.(figure 5) Quartz, which is mainly silica, and mica are two of the minerals that are dispersed when granite is weathered and breaks down. Like kaolinite, china stone is formed from the decomposition of granite. It is often known as "rotted granite". When granite is fully rotted, it forms kaolinite.[4] China stone, therefore, has the same origins as china clay but it is a slightly different combination of minerals. Its structure, however, is very similar. It is crystalline and platey in texture but the platelets are quite large compared to those of china clay. China stone can, however, be ground up to form a fine, platey material with a small particle size and thus can become a fairly plastic material. China stone is however less plastic than china clay but china clay can be added to china stone to make it more plastic. This process was achieved progressively at the imperial kilns at Jingdezhen in south China where in time, gradually more and more china clay was added to the china stone body, thus increasing its plasticity. The limitations of pure china stone bodies are reflected most clearly in Dehua wares from Fujian province in south China.(figure 60) These wares, commonly known as "*blanc de chine*", have bodies made of pure china stone and thus are very hard and sugary in texture. This body is quite well suited to moulding and these kilns are famous for their Buddhist figure models. However, because it is not very plastic, this body is not well suited to forming anything on the potter's wheel other than very small or very thick vessels. One point that should be noted with regard to plasticity is that very plastic clays tend to shrink a lot during drying and firing. As discussed above, the smaller the particles in a clay body, the more plastic the clay is because it holds more water. The more water that exists in a clay, the more there is to be removed and thus the more it shrinks. China clay, therefore, exhibits quite high shrinkage, whereas china stone, with its larger particle size, has lower shrinkage. This concept will be discussed more fully in chapter five in respect to firing. As we have seen, primary clays have larger particles than secondary clays and thus are less plastic and shrink less.

Clays require much preparation before they are suitable for use as ceramic bodies. China clay, for example, requires quarrying and then much levigation or sieving and removal of large inclusions before a pure white sediment is formed which is then mixed with water. This is then a suitable body material because without preparation, bits of rock, sand, fossils and other impurities would be left in the clay. Some of these inclusions are quite useful however.One example is unweathered feldspar which is useful as a glass-former or hardener in ceramic bodies and glazes. China stone also requires much preparation before it can be used to form ceramics. It too must be quarried but then it must be crushed or pulverized to reduce the particle size, then levigated and left to dry slightly before it can be formed into bricks. It is these bricks that give it its other name "petuntse" or *baidunzi* which means "little white bricks" in Chinese.

These bricks are then crushed and mixed with water to form a type of clay. China stone deposits often contain some kaolinite naturally, which helps to increase its plasticity. As noted above, china clay is higher in alumina than china stone which makes it higher firing. It matures at a

---

[4] Cuff, *Ceramic Technology for Potters and Sculptors*, p. 40.

higher temperature. The term used to describe high firing clays is "refractory". Thus, northern Chinese ceramics are made from highly refractory clays and southern chinese ceramics are lower firing or "fusible" as they are higher in silica. The impact on visual appearance of vessels made from these materials will be discussed in chapters two and five.

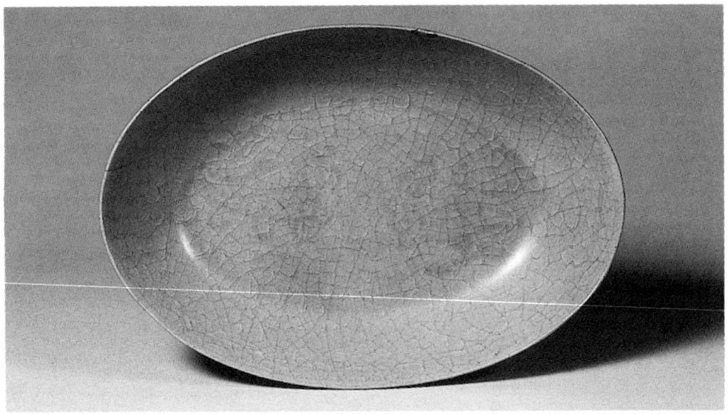

2. Ru ware brush washer, Northern Song dynasty, early 12th century. PDF 76.

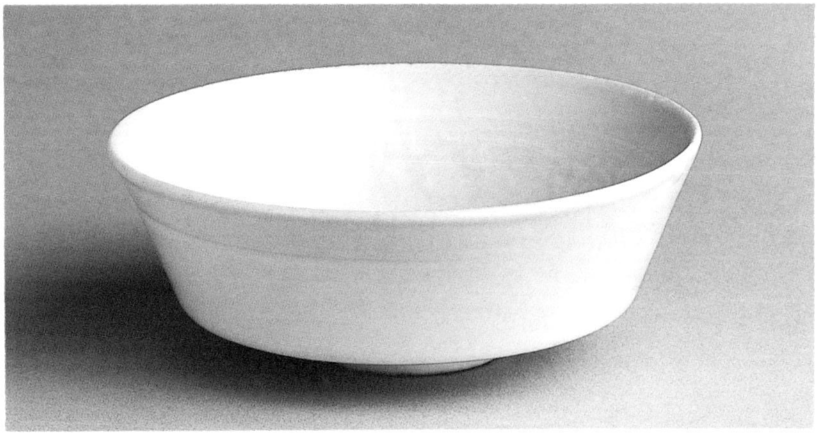

3. Shufu ware bowl, Yuan dynasty, 1st half of 14th century. PDF A455.

2. Glazes

Most ceramics are glazed, either for decoration or for "sealing", to render them impervious to water, or both. The exception to this is terracotta flower pots and other earthenwares which must be porous by necessity. In China, the earliest glazed ceramics appear to have been made in the Shang dynasty (1700-1027BC). Some earlier glazes are probably the result of ash falling on the pots in the kiln which gives a glaze-like effect. In the Shang dynasty, the effect was produced deliberately and this transformed ceramic production. Prior to the invention of glazes, ceramics were generally low-fired so fairly soft and porous and they were painted with unfired pigments. Glazed ceramics were much more useful and more decorative.

Glazes can be transparent or opaque, brightly coloured or subtle, shiny or matte. Like clays, glazes come in a wide range of colours and types but their basic chemistry is quite simple. A glaze is essentially a layer of glass on the surface of a vessel.(see figure 20) It is made from the same ingredients as the ceramic body but with more "flux", a substance that makes the glaze or body material melt at a reasonable temperature, and more "glass-formers" such as quartz or feldspar. As we have seen in chapter one, in order for clay to become "ceramic" it must be baked or fired so that water is removed and the clay particles fuse together. In the case of porcelain, the components of a clay body must melt sufficiently to form a type of glass which makes the body hard. Similarly, a glaze must melt so that it can flow over the vessel and coat it and then cool into a hard glass covering. If fluxes were not present in or added to bodies or glazes, different kilns would have to be invented because they would have to be fired to 1700°C, the temperature at which silica, one of the main ingredients in ceramics, melts. The most sophisticated kilns used in pre-modern China could reach a maximum temperature of about 1400°C. It is the fluxes, therefore, that make glazing possible.

Glazes are generally classified according to what kind of flux is present. In Chinese ceramics we are concerned with four main types of glazes: lead, lime, lime-alkali and porcelain or acid-rock glazes.[5] To a small extent, alkaline glazes were also used in China, principly on *fahua* wares (see figure 38) or any turquoise glazed vessels.(see figure 19) Lead glazes are those glazes which contain lead as their flux. They are low firing glazes that are fired in an oxidizing atmosphere (clean, not smoky--see chapter five) and are characterized by brilliant colours. Lead glazes were used extensively in China by the Han dynasty (206 BC to 220 AD). The most famous lead glazed wares in Chinese ceramics are the *sancai* or three-colour wares. (figure 13) These wares were made in the Tang dynasty (618-906 AD) in north China and their glazes feature all of the colours then possible in a lead glaze: green, blue, amber, cream, brown, yellow and black.

---

[5]Wood, N., in Oriental Ceramics Society, *Iron in the Fire*, p. 12.

They are usually green, yellow and brown or amber. (Note: The Chinese use the term *sancai* to describe any ceramics decorated with the "three-colour" palette. This includes the large Ming Buddhist figures and enamel-on-biscuit wares as well as Qing porcelains utilizing this palette.)

Glazes are coloured with mineral pigments and the most commonly used pigments in Chinese ceramics are: iron, copper and cobalt (see figures 17-19). *Sancai* wares feature glazes coloured with all three pigments. Yellow, amber, brown and black are produced from iron pigments. The green and turquoise are from copper pigments and the dark blue from cobalt. The turquoise glaze is used infrequently in Tang ceramics and is the result of copper in an alkaline glaze or a glaze with soda or potassia as its flux. Turquoise glazes were more common in Islamic and Near Eastern ceramics where alkaline glazes were the norm. In addition to green and turquoise, copper can also be used to produce red colours in a high temperature glaze (see figure 19) and the earliest use of copper red glazes thus far is in Changsha ceramics of the Tang dynasty. Changsha wares are southern wares from Hunan province and are known for their innovative glazes and decorative techniques. Whereas copper green glazes are lead fluxed or porcelain glaze and fired in oxidation, to produce a red glaze from copper, a reducing atmosphere is usually necessary. Some Changsha wares feature red and green colours together and it is not known how this effect was achieved. It has been speculated that this effect could be achieved by altering the melting point of the red glaze so that it melts at a higher temperature and it thus can be reduced after the green glaze has already melted. The process could also be reversed so that the red glaze reduces then melts and then the pot is oxidized just before the green glaze melts. Alternatively, it is possible that a reducing agent was added to the red glaze which is effective in any type of atmosphere or that the red glaze was added after the vessel was first fired in oxidation then it was fired a second time in a reducing atmosphere.[6] As we can see with *sancai* wares, the iron pigment can be used to produce a wide range of colours. Iron is perhaps the most versatile pigment used in ceramics. It can produce a number of colours in both oxidizing and reducing atmospheres. In general, the iron colours produced in an oxidizing atmosphere tend to be "warm" colours: yellow, brown, rust, amber or black.(see figures 17, 25, 26) Iron fired in reduction produces "cool" colours such as green or blue.(see figures 14, 15, 27) Iron can also be used in any type of glaze and thus glazes coloured with iron are found in Chinese ceramics of all periods including: Han lead glazes, Tang *sancai* wares, Song celadons from north and south China, *qingbai* or *yingqing* wares and later Jingdezhen porcelains, both monochromes and enameled wares.

Another early type of glaze is the lime glaze used by the Chinese potters. These glazes have high amounts of calcia or lime for their flux. This is added to the glaze in the form of wood ash.

---

[6] Wood, Nigel in Scott, ed., *Chinese Copper Red Wares*, PDF Monograph Series No. 3, pp. 15-16.

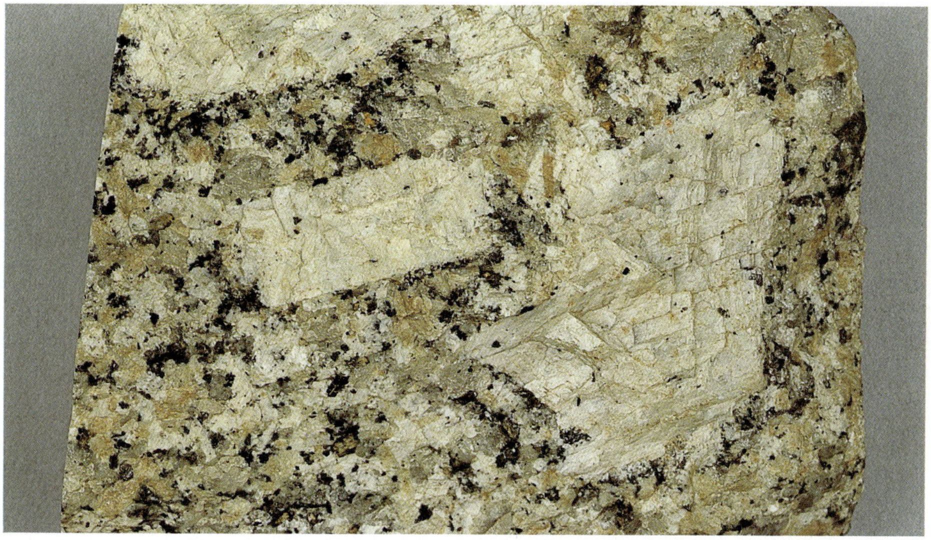

4. Granite, UCL, Department of Geology Petrology Collection. P55. This sample shows large white feldspar inclusions.

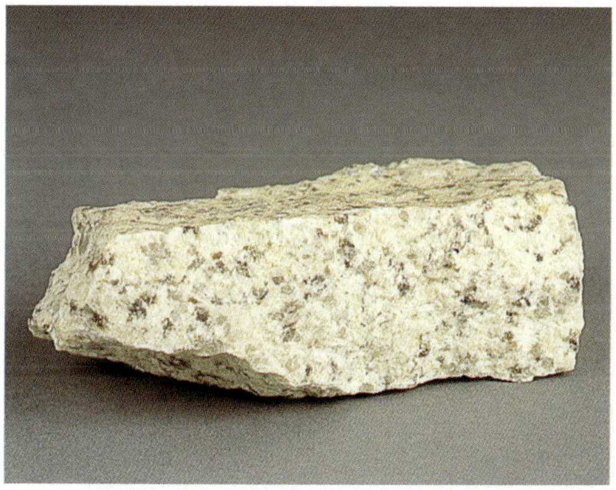

5.Chinastone, UCL Department of Geology Petrology Collection. P4238.

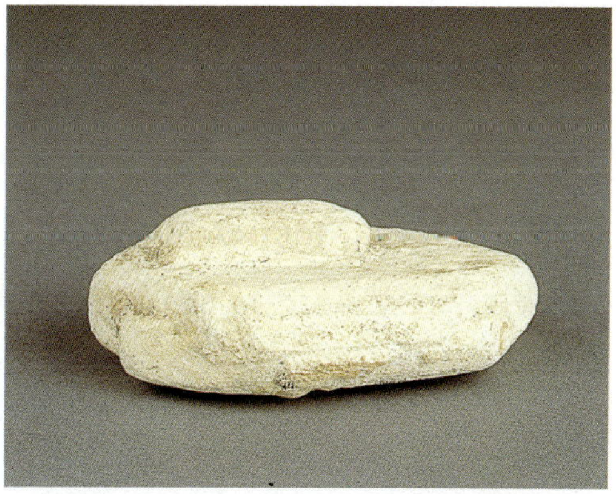

6. Kaolinite, UCL Department of Geology Petrology Collection. 88:1(2).

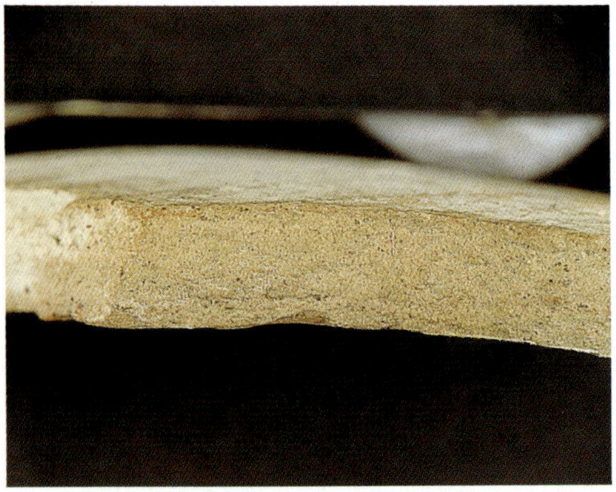

7. Earthenware body, PDF Study Collection shard.

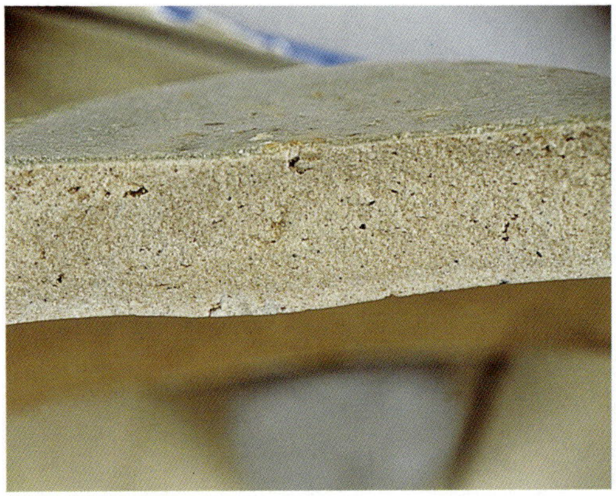

8. Stoneware body, PDF Study Collection shard.

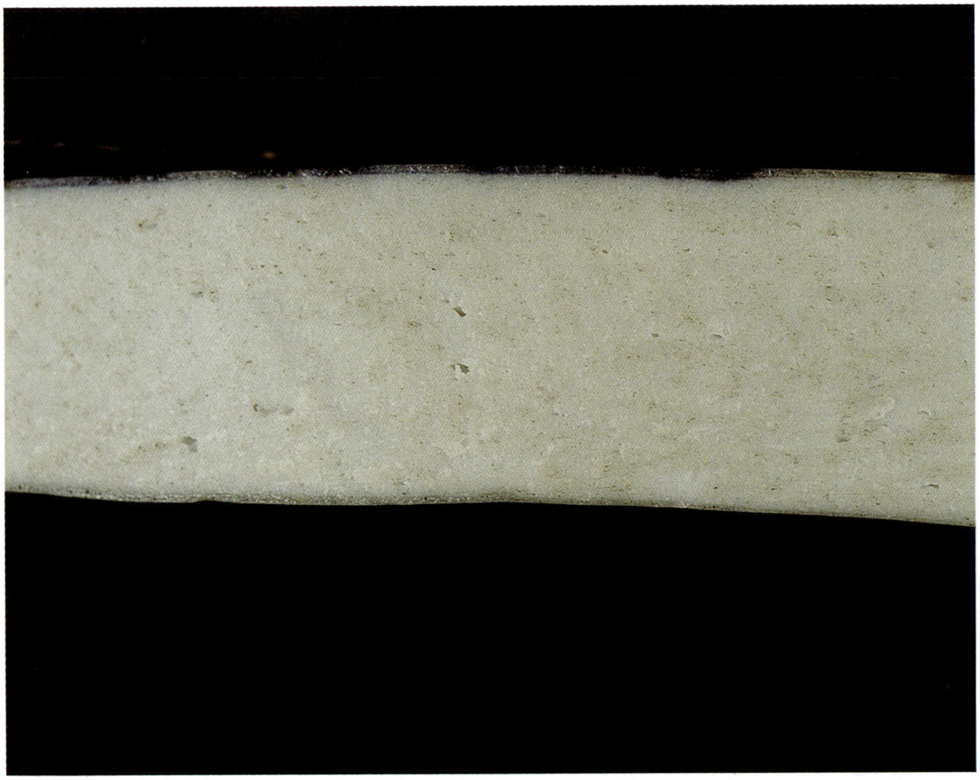

9. Porcelain body, 15th century, PDF Study Collection shard.

Therefore these glazes are often called ash glazes. The most famous example of a lime glaze in Chinese ceramics is perhaps the Yue celadon glaze. (figure 14) Yue wares were made in Zhejiang province until the 10th century AD. They are classified as southern wares and have typically southern china stone bodies. The Yue glaze is coloured with iron fired in reduction and contains on average 15-17% calcia. Lime glazes are characterized by their thinness and their "frostiness". They must be thin because the high amounts of lime flux make them very fluid and thus there is a danger of them running off the pot if too much glaze is used. The frostiness is caused by minute lime-rich crystals which appear in these glazes during cooling. (Note: this process of crystal formation in glazes is often called "devitrification". See chapter five for full explanation) Therefore, these glazes tend to be slightly matt and can appear sugary in texture if high in silica. Unlike the lead glaze, this is a high firing glaze and in the case of Yue wares, they were probably fired to 1300°C. High fired glazes first appeared in China in the Shang Dynasty and are one of the most important achievements of Chinese ceramics. The earliest lime glazes are much higher in calcia than the later examples in the Song dynasty (960-1279 AD).

A particular aspect of Chinese ceramics is that in general they are "green glazed", or glazed before being fired and then fired only once, except in the case of overglaze decoration. Most other ceramic traditions, including contemporary art or studio ceramics feature vessels that have to be biscuit fired, or fired once to harden the body then glazed and fired again. This has not been necessary in Chinese ceramics because the glazes and bodies used are very similar in composition and thus there is good "glaze fit". Glaze fit is important because when a pot is being fired the body and the glaze are expanding and contracting. If expansion and contraction occur at different rates, then the glaze will crack or peel off the vessel. In Chinese ceramics, the similar compositions of the glazes and bodies in most wares ensure that this does not occur. This is particularly noticeable in Chinese porcelains of the Ming and Qing dynasties where there appears to be almost no difference between the glaze and body.(see figure 9) However, Chinese potters were nothing if not inventive and we shall see below how manipulation of glazes and bodies can produce spectacular effects.

The ceramics of the Sui and Tang periods (581-906 AD) in China are some of the most versatile wares ever produced. In this period both low firing and high firing glazes were used and porcelains were produced for the first time. Lead glazes, lime glazes and early lime-alkali glazes were all used in this period. The lead glazes used on *sancai* wares were usually applied to low firing, earthenware bodies and thus the vessels need only be fired once. However, some Tang ceramics have white firing bodies that are higher firing than the usual earthenware body. Thus a low firing lead glaze would have been applied after first biscuit firing the vessel, which would then have been re-fired in a second, low temperature firing. This is not necessary with high firing lime and lime-alkali glazes as they are applied to high firing bodies. The Yue ware body, incidentally, consists of almost pure china stone and thus was quite high firing. This was not a porcelain body however because by Western criteria, it must be white and translucent to be considered porcelain. In the North, a true porcelain was produced at the Xing and Gongxian kilns in the Tang dynasty (618-906) (see map). The Xing wares had lime glazes and the Gongxian potters developed the earliest lime-alkali glaze to be found thus far in Chinese ceramics. The Gongxian kilns are perhaps best known for using underglaze cobalt blue decoration, the first kilns to do so in the history of Chinese ceramics. The lime-alkali glaze is one of the most beautiful glazes produced in Chinese ceramics and one of the most versatile. It is high firing and contains both lime and potassia (an alkali) as fluxes. This type of glaze is most often associated with Song ceramics and is characterized by the smooth, unctuous celadon glaze found on Longquan wares, a southern ceramic produced in Zhejiang province from the Song to Ming dynasties.(figure 15) The Longquan glaze typically contains 6-12% calcia and 3-6% potassia. As with all celadons, this glaze is coloured with iron and is fired in reduction. (Note: "Celadon" is a western term. The Chinese refer to ceramics with celadon glazes as "greenwares" or *qingci*.) The Longquan ware body is composed mainly of china stone and thus is fairly high firing. Lime-alkali glazes contain more flux than lime glazes and therefore melt at a lower temperature. One of their special qualities is the potential for "unctuousness" which can be achieved if these glazes are slightly underfired. The Longquan glaze, for example, was typically fired to around 1200°C, whereas the glaze "matures" or fully develops at about 1250°C. Another Song celadon that features a lime-alkali glaze is the northern Yaozhou celadon.(figure 29) This ware is characterized by a thick, glossy, transparent glaze that is usually slightly olive green in colour. This colour is the result of iron fired in reduction but with a slight titanium impurity that turns glazes and bodies slightly yellowish. As a northern ceramic, Yaozhou ware has a clay-rich body and is thus higher firing than Longquan ware. The Yaozhou glaze is fully fired, and thus transparent, not slightly opaque like the Longquan glaze. The lime-alkali glaze can therefore be quite versatile and has been used continuously in Chinese ceramics since the Tang dynasty (618-906).

The final glaze type used in China is the porcelain glaze or acid-rock glaze. It is so called because it has quite a high proportion of china stone in the glaze and china stone is an acid rock. From the invention of true porcelain in the 7th century, a number of different porcelain glazes were used on Chinese ceramics. As we have seen, the earliest were lime glazes which were used on northern porcelains. In the south, however, a different porcelain glaze, based on china stone, was used and it was this glaze that developed into the sophisticated glaze used on imperial porcelains at Jingdezhen in Jiangxi province. The earliest porcelain glaze used in the south can be found on *qingbai* or *yingqing* wares which were made at Jingdezhen and other areas from the Song period onwards (see map and figure 31). The *qingbai* glaze typically consisted of 65% porcelain stone and 35% glaze ash which is mostly lime. This is therefore a high lime glaze but the small amount of alkali found naturally in china stone makes the glaze closer to a lime-alkali glaze than the early northern porcelain glaze. Because of the high amount of flux, this glaze is very fluid. Tiny amounts of iron in the glaze render it a pale, watery blue. The next porcelain glaze developed at Jingdezhen was the *shufu* glaze in the Yuan dynasty (1279-1368). This glaze is also based on china stone but contains much less flux than the *qingbai* glaze, approximately 5-6% calcia. This made the glaze higher firing and the body had to be adjusted accordingly. Unlike the *qingbai* glaze, the *shufu* glaze is rather opaque and quite thick.(figure 3) When underglaze blue decoration began to be used extensively at Jingdezhen in the Yuan dynasty (1279-1368 AD), it seemed that neither of these two porcelain glazes was suitable. The *qingbai* glaze was too fluid and caused the blue to run and the *shufu* glaze was too stiff and opaque so that the blue did not show through. Potters therefore developed a new glaze that was intermediate between the *qingbai* and *shufu* glazes and more like a classic lime-alkali glaze. The glaze developed for porcelain decoration with underglaze blue contained about 8-10% calcia. It was essentially a colourless glaze, though porcelain glazes of this type are often erroneously described as "white" glazes. In later periods at Jingdezhen, a true white glaze was developed which obscured underglaze painting but was perfect for overglaze enamel decoration (see chapter three). This glaze is similar to the *shufu* glaze and typically contained 5% calcia and 7% alkalis and was somewhat opaque due to undissolved silica in the glaze. (Note: china stone is very high in silica so a glaze based on it would be high in silica as well. In the white porcelain glaze, there is not enough flux to ensure that all of the silica in the glaze is melted when the body is fired to maturing temperature.) It differed from the early lime-alkali glaze used at Gongxian in the Tang dynasty in that it was based on china stone not china clay.

Some of the more spectacular and technically sophisticated glazes used in China fall outside the range of glaze types discussed above. These include: Jun, Guan, Jian or *temmoku* and copper red monochrome glazes.(see cover illustration and figures 21, 10, 19) The Jun, Jian and copper red monochrome glazes are all variations on lime-alkali glazes but with a particular glaze chemistry that encourages the glaze effect known as "liquid-liquid phase separation". In each of these glazes, phase separation is responsible for giving the glaze a certain colour or texture that cannot be achieved by conventional means. By definition, phase separation occurs when a liquid that is homogenous at high temperatures separates into two or more liquid phases on cooling. A "phase" is simply another name for a state of matter, eg. solid, liquid or gas. In glazes, the phase separation is actually a "liquid-liquid" phase separation in that one liquid splits into two or more liquids or liquid phases. Glass is classified as a liquid, a "supercooled" liquid to be exact, and therefore so are glazes. In practice this means that in parts of the glazes that feature phase separation, the glaze contains droplets of glass that are suspended in the main glaze or glass matrix. These droplets are extremely small, of "colloidal" size, and interfere with light passing through the glaze. In the Jun glaze, phase separation is responsible for its characteristic opalescent blue colour because the droplets are so small that they only reflect blue light, which is what we see when we look at a Jun vessel. All other light waves pass through the glaze. Jun ware also has another feature which makes the glaze appear especially bright. At the area where the glaze meets the body, the "glaze-body interface", there is a thin layer of white crystals that reflect light back through the glaze and enhance its beauty to the viewer. (see figure 20) This is a layer of anorthite (lime feldspar) crystals which develop during the long slow cooling process which is a feature of Jun ware production. This process of crystal growth is known as devitrification and will be discussed in greater detail in chapter five. Incidentally, the blue "splashes" found on Tang "Jun" wares are phase-separated as are the spots and splashes found on Song Jizhou tortoise shell wares.(see figures 22, 23) (see chapter three for discussion)

While phase separation in Jun glazes is fairly well understood, its appearance in Jian and copper red monochromes glazes is not. Jian wares were produced in Fujian province in the Song dynasty and are characterized by a very dark brown body and a black glaze with rust coloured streaks on the surface.(figure 10) The glaze is overloaded with iron (ca. 6%) and the excess precipitates out to form these streaks or sometimes large droplets in the case of "oil spot" glazes.(figure 11) Only a certain amount of iron can be dissolved in a glaze (up to about 5.5% in lime-based glazes) and therefore any excess will likely be in suspension. In the case of Jian glazes, it is possible that the formation of the little glass droplets during phase separation helps to carry the excess iron in the glaze to the surface where the bubbles can burst and thus cause the streaking that is described as *tu hao ban* in Chinese or "hare's fur". Before this glaze reaches the streaking stage, it passes through a phase in which there are black iron specks on a light brown or greenish base glaze. If stopped at this stage, the glaze becomes the "tea dust" glaze that is familiar in Chinese ceramics.(figure 12) It then moves on to the "oil spot" stage and then to the "hare's fur" stage where the spots burst and run down the sides of the vessel. Black glazes have been produced in China since at least the third-fourth century AD but it is the Jian "*temmoku*" glaze that is perhaps the most sophisticated example. The earliest black glazes were the high lime glazes used on Deqing black wares from Zhejiang province. These black glazes were not very successful because the high-lime base glaze cannot support more than 3-4% iron and the Deqing black glaze contained ca.6.5%.[7] The lime-alkali glazes used in the Tang and Song dynasties produced better black glazes. A black iron glaze is sometimes called an "iron solution" glaze because all the iron is dissolved or in solution in the glaze. A red iron glaze is an "iron suspension" glaze because there is too much iron in the glaze so some of it remains undissolved, in suspension. Thus red iron glazes contain more iron (up to 15%) than black iron glazes!

The copper red glaze is still something of a mystery. It is essentially a standard porcelain glaze that is coloured with copper and fired in reduction. However, this glaze colour is extremely difficult to control. Copper volatilizes quite easily and thus many copper reds turn grey or in some instances the colour even disappears altogether. In the best copper red monochromes, particularly those of the 15th century with a bright, almost cherry red colour, it appears that the colour is actually a very thin coloured layer sandwiched between two colourless layers.(see figure 19) The copper particles must be very small to give a good red and not large and clumped together which one sees on the examples with a slightly liverish colour. This is particularly

[7]Wood, N., *Iron in the Fire*, The Oriental Ceramic Society, 1988, p. 12-13.

noticeable on copper red monochrome glazes of the Hongwu period (1368-98). To stop the copper particles from clumping together, the glaze has to be fairly fluid but not so fluid that the copper reaches the surface and volatilizes or burns away. The normal 15th century porcelain glaze is fairly stiff, so it must have been altered to be used as a copper red monochrome glaze. It is possible that this was achieved by slight overfiring of the glaze[8], which would render a stiff or viscous glaze more fluid. Phase separation appears in this glaze but it is not known what effect it has on the glaze. It should be noted that this glaze is a lime-alkali glaze and it is these glazes that most often exhibit phase separation. The 15th century copper red monochrome glaze is referred to as "fresh red" or *xianhong* in Chinese. It was copied in the 18th century and these later examples are often called "Lang yao" or "sang de boeuf". These terms are often used to denote a particular hue but "Lang yao" should only be used to describe those copper red monochromes that were produced during the time in which Lang Tingji was the supervisor of the imperial kilns (1705-12).

The final unusual glaze used in Chinese ceramics that will be discussed here is the Guan glaze. (figure 21) Guan ware was the Southern Song (1127-1279) official ware ("guan" means official in Chinese) that was produced for the court in and around Hangzhou.(see map) Its glaze is characterized by a heavy crackle and is a variation on a lime glaze. It is higher in alumina and lower in silica than the usual lime glaze and tends to be slightly opaque or translucent. The cracking of the glaze is deliberate which is why it is called "crackle" rather than "crazing" which is a fault in the glaze. The crackle is caused by the deliberate underfiring of a glaze which is low in silica. Silica prevents glaze crazing which occurs because the glaze does not fit the body.[9] The body expands during the firing as it heats up and in this case the glaze is not expanding with it, so when the glaze begins to cool it cracks to relieve the pressure. When it first cracks, the cracks are widely spaced and this is known as "primary crackle". If the cracking is allowed to continue, very small, closely spaced cracks will emerge that are described as "secondary crackle". This is much like wearing a garment that is too small and then stretching, causing the fabric to rip. A Guan glaze fired to maturity does not craze and looks much like an ordinary celadon. Because this glaze is slightly underfired, it is somewhat opaque as the ingredients in the glaze have not melted fully and thousands of gas bubbles could not escape or pass through the viscous glaze. In addition, the extra alumina in the glaze encourages the growth of lime

---

[8]Wood, Nigel in Scott, ed., *Chinese Copper Red Wares*, PDF Monograph Series No.3, pp. 25-26.

[9]Crazing or crackle can be elimiated by adding more silica to the glaze mix which decreases the proportion of alkalis in the glaze or by altering the body composition and firing to a higher temperature which encurage the body to shrink. See both Hamer and Cuff for full discussion.

feldspar (anorthite) crystals which also bend and scatter light, lending an opaque appearance. Guan ware is a rather controversial ware as a variety of Guan-type glazes appear to have been produced, ranging from the lustrous blue wares with wide, pale (unstained) crackle traditionally attributed to the Xiuneisi kiln site (as yet unexcavated) to the yellowish, more densely crackled wares included among the finds from the Hangzhou Jiaotanxia kiln site.(figure 21) Some of the better quality examples of Guan ware appear to have several layers of glaze and each layer was probably fired before the next layer was applied, a most unusual practice in Chinese ceramics. Thus the best examples of Guan ware have very thick glazes on a thin, dark body. The glaze is noticeably thicker than the body. Incidentally, crackle in glazes is often stained for emphasis and this is best done when the vessel is still warm so the cracks are still open. Once the glaze cools and smoothes out, the cracks close and staining would be much more difficult. Charcoal ash is a most effective stain. There is another type of ware with a crackled glaze that should be mentioned. Ge ware is a ware that is often referred to in catalogues of Chinese ceramics and many museums describe certain of their "crackled" wares as Ge ware but to date there are no concrete criteria for differentiating this ware. Traditionally, Ge ware is the term used to describe some wares with a densely crackled glaze that date to the Song or Yuan period. Examples have been found in Yuan tombs but not Song tombs. No kiln site has been discovered yet, however there are various written records from the Yuan dynasty onward that mention Ge ware but usually in conjunction with Guan ware or Longquan ware. In any case Ge ware is included among the "five great wares" of the Song dynasty along with Ding, Ru, Jun and Guan. Currently, there are three theories about Ge ware: it is the name given to Yuan period Guan ware; it is a separate ware in its own right or it is the crackled ware produced at Longquan. Because there is so much confusion about Ge and Guan ware, it is perhaps safer to refer to all southern crackled wares of the Song and Yuan periods as "Guan-type" wares, as the technology is inherently the same.

All of the above mentioned "unusual" glazes belong to the category of monochrome glazes which are an important part of Chinese ceramic production. Glazes of just one colour have been produced since the very first glazes were invented in China and almost every kiln in Chinese history produced some type of monochrome glaze. Some of the most familiar monochrome glazes however are those produced on imperial porcelains of the Yuan, Ming and Qing dynasties. Among the earlier high-fired monochrome glazes were those produced on Song Ding wares including black and brown monochromes.(figure 26) Also in the Song dynasty, high-fired copper green and red monochrome glazes were used on porcelain at the Rongxian kilns in Guangxi province. In the Yuan dynasty, turquoise and cobalt blue monochromes were popular but it was in the Ming dynasty that the monochrome glaze palette increased dramatically. This is partly due to the fact that in 1369, the emperor Hongwu decreed that the ritual vessels used for imperial ceremonies had to be monochrome glazed porcelains of particular auspicious colours: red for the altar of the sun, yellow for the earth, blue for heaven and white for the moon. The red monochrome glaze is a reduced copper red glaze (figure 19), the yellow ("imperial yellow")is a low-fired iron yellow enamel (figure 25), the blue is a high-fired cobalt blue glaze (figure 18) and the white is a regular porcelain glaze. Other monochromes used in the Ming dynasty include high-fired iron brown or "*café-au-lait*"(figure 17), turquoise (figure 19), copper green and iron red on *kinrande* wares, aubergine (from manganese) as well as archaistic copies of Song celadon glazes which are similarly made from small amounts of iron fired in reduction. There are a few very rare examples of black monochrome glazes on Ming porcelains but this glaze is more common in the Qing dynasty, where it is made from a combination of iron, cobalt and manganese which produces a "mirror black" glaze.

In the 18th century, new monochrome colours emerged which developed from the new overglaze enamel palette that included both opaque colours and pink. Some of these new colours were adapted for monochrome glazes and produced colours that were not possible in the Ming dynasty including: rose pink (colloidal gold) and lemon yellow (lead-stannate or antimonate). Other popular Qing monochrome colours include: lime green, gold, dark purple, pale turquoise and coral red. The most unusual Qing monochrome colour is related to a monochrome that was produced in the Ming dynasty but its technology is still something of a mystery today: the "peachbloom" glaze.(figure 24) This glaze is made from copper and features both copper green and copper red in the same glaze. It became popular during the Kangxi period (1662-1722) and is still widely admired by both collectors and potters. It is sometimes called "drunken beauty" or "bean red" in Chinese. To achieve this unusual effect, some potters believe that copper pigment was sprayed on to a clear glaze and covered with another clear glaze so that it was effectively sandwiched between two clear glazes. As the pigment was sprayed on, some areas would have more pigment than others and in these areas the copper would seep through the top layer of glaze and re-oxidize or turn green in the firing. The other areas with less pigment would be dissolved in the glaze and thus would reduce to a nice copper red colour. This glaze is a high lime glaze (ca.17%) and analyses show that a large amount of copper was used for pigment (ca.2.4%).[10]

[10]Wood, N., *Chinese Copper Red Wares*, PDF Monograph No. 3, p. 28-29.

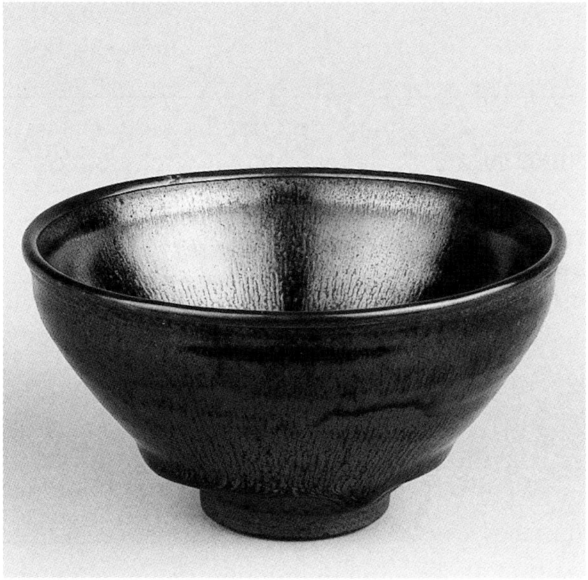

10. Jian ware bowl, Song dynasty, 11-13th century. PDF 303.

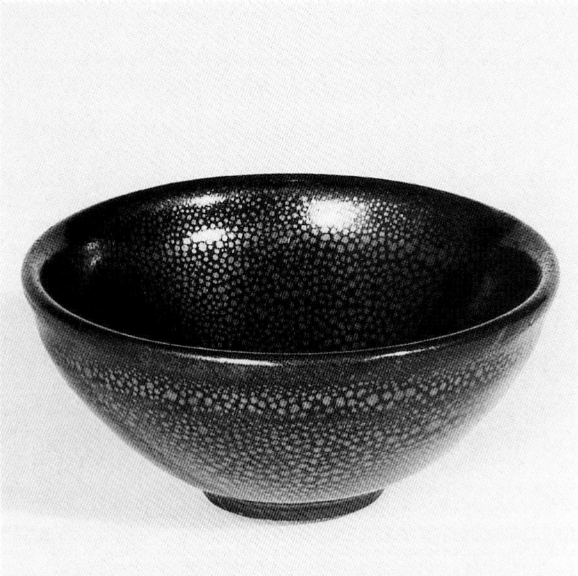

11. Bowl with "oil spot" glaze, Northern Song/Jin dynasties, 11th-12th centuries. PDF 301.

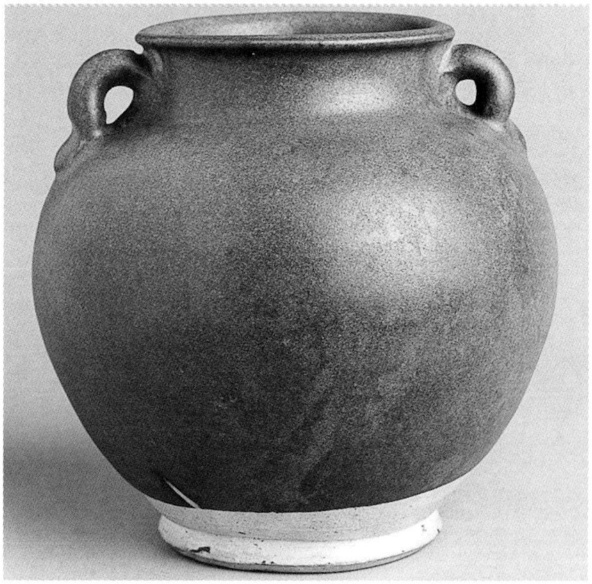

12. Jar with "tea dust" glaze, Tang dynasty 8th-9th centuries. PDF 330.

3. Decorative Techniques and Forming Methods

As we have seen in Chapter One, clay differs from mud and other earthy materials in that it is plastic--that is, it can be formed into a shape and will hold that shape while it is fired. To make ceramic forms, clay can be shaped in three ways: by hand (hand built), on a potter's wheel (thrown) or by moulding. In Chinese ceramics, the wheel is the most commonly used method. The potter's wheel is a rotating disc on top of a fly wheel and has been used in China since the Neolithic period. It produces characteristically round shapes like bowls or vases. The action of throwing clay on the wheel involves centering a lump of clay on the turntable and then gradually hollowing out and raising the clay with the fingers until the desired form is created. This process leaves tell-tale "throwing rings" on vessels which can be smoothed out but are often visible on finished pots. The potter's wheel can be used to create very small or very large shapes. As the potter is limited by the size of wheel and weight of clay he can manage, very large shapes are often thrown in several parts which are then joined together by "luting" which involves wetting the surfaces to be joined and scoring them so that they stick together then smoothing over the join lines where the surfaces meet. A very large wine jar, for example, can be made by throwing two large bowl forms then inverting the top one and joining together at the rims.(see fig 41) Once a vessel form has been thrown, the pot needs some finishing to trim it, thin it out or smooth it. This finishing process is known as "turning" because it is usually done on the wheel while it is turning slowly.

Moulding and hand building were also used in China but to a lesser extent than throwing. Moulded forms often have very elaborate shapes, which cannot be thrown, or complex impressed decoration.(see figure 32) When a moulded piece is made, a basic shape is usually thrown on the wheel and then this piece is placed over a slightly porous, fired ceramic mould.(figure 35) It is pressed on or beaten on with a paddle and paddle marks can often be seen on fired moulded vessels. The porosity of the mould is important because it means that the wet clay placed on it will be absorbed slightly thus giving a nice sharp impression. It is not so porous, however, that the clay will be too difficult to remove from the mould. There is some evidence that double moulds (inside and outside) were used in some Chinese ceramics but these were the exception. Another type of moulding was used more recently in China and the West when large quantities of similar wares were being produced. This type of mould is known as a "jigger" or "jolly" today and it is basically a mechanical moulding process in which clay is placed on or in a spinning mould and a profile is pressed down on it to form the other surface. A jigger forms the inside of a plate and a jolly the outside. This type of moulding is quite a late invention and was preceded by a similar process which involved using a fast wheel and a template. Here a mould was placed on a fast moving potter's wheel and then a disc of clay was placed on it. While the potters' wheel is moving a template or profile was pressed down on the clay to shape the form.

Moulds were also used for decorating Chinese ceramics and this will be discussed below in the section on decorative techniques. Some of the most complicated ceramic forms in China were created in moulds, including the famous Dehua or *blanc de chine* Buddhist figures.(figure 60) The basic forms of these were usually made by pressing clay into two mould halves and then joining them together and smoothing the join before firing. In Neolithic China, hand building was also used to form pots. The most common method employed was that of coiling in which long rolls of clay are gradually coiled up on top of each other until a vertical form is created. When one roll of clay ends, another one is pressed on to it so there are numerous joins that need to be smoothed over in the finishing stage. This smoothing is usually done by beating the coiled pot with a paddle which often leaves visible flattened areas on the fired pots. This method was somewhat laborious and was superceded by the wheel which seems to have first appeared about 3000 BC.

underglaze decoration

Once a clay vessel has been formed and turned or finished, it is often decorated before glazing. This can be done in numerous ways but in China, the decorative techniques used can be grouped into three main categories: painted decoration, *intaglio* decoration and relief decoration. Decoration applied after glazing will be considered in the next section. The most common and the oldest form of decoration on Chinese ceramics is painted decoration. Some early ceramics produced in China were often highly decorated with paints made from mineral pigments that were applied to the fired vessel in the form of a slip or runny clay mixture. This type of painted decoration was not very permanent because it was not fired or melted on. When first painted, these wares were probably quite bright and colourful, however time and burial has faded the pigments considerably. Unfired painted decoration continued to be used in China in the Han dynasty (206 BC - 220 AD) on vessels and even later on figures made for burial. Ideally, decoration on ceramics should be more permanent and thus fired-on pigments began to be used from about the 3rd century AD. These fired-on paints are applied to the dried but not fired vessel which is then glazed and fired. Thus, this type of painting is known as "underglaze painting". Underglaze painting is an extremely effective way of decorating pots and is done by mixing mineral pigments with an aqueous medium, usually water, which could then be used like a paint. The earliest underglaze painting on Chinese ceramics was made with iron pigments and can be found on some early greenwares from Zhejiang province.[11] Underglaze iron continued to be the main type of underglaze painting until the Tang dynasty (618-906 AD) when other pigments such as copper and cobalt were used for underglaze painting.(see illustration on the back cover) The earliest underglaze copper painting thus far appears on ceramics from

---

[11]See the exhibition catalogue from the Fung Ping Shan Museum in Hong Kong, *Greenwares from Zhejiang* which is listed in the bibliography.

Changsha in Hunan province. Here we see both underglaze copper green and underglaze copper red being used. As we have seen in the chapter on glazes, copper is an extremely difficult pigment to work with but quite versatile. It can be used to create several different colours, depending on the base glaze and the kiln atmosphere. Copper green is usually produced in an oxidizing atmosphere whereas copper red is produced in a reducing atmosphere. In Changsha wares, the two colours often appear together. (See chapter two for full discussion) For underglaze painting on porcelain, copper red is the more usual variant. This type of painting can be seen on a range of Chinese ceramics including: *qingbai* wares, Jindezhen imperial porcelains and Jun wares as an in-glaze pigment (see discussion below). When used on porcelain, the vessel is made and dried, then painted with copper red paint, then glazed and then fired. As noted in chapter two, on some porcelains, the underglaze red painting appears greyish or disappears due to the volatility of the pigment.

Underglaze copper red is often combined with the most expensive underglaze pigment: cobalt blue.(see back cover illustration) The first use of underglaze blue painting on porcelain known to date appears on Gongxian porcelains of the Tang dynasty (618-906 AD). These early examples were probably made for Near Eastern patrons and the cobalt used was probably imported from the West. Underglaze blue painting began to be used more extensively on imperial ceramics at Jingdezhen in the Yuan dynasty.These wares are usually called "blue and white porcelains". The cobalt used at this time was also probably imported from the Near East. Imported cobalt was high in iron and tended to have blackish or brown specks in it which can be seen most clearly on blue and white porcelains of the Xuande period (1426-35)(see back cover illustration). This effect is called "heaping and piling" by the Chinese and was considered desirable, so much so that it was imitated on Qing blue and white wares decorated in the Ming style. At various times in the Ming dynasty, local Chinese cobalt was used and thus a different tone of blue can be seen on some blue and white wares. For example, on blue and white wares from the mid-Chenghua period (1465-87), the blue decoration is usually quite pale. Local Chinese cobalt tends to be high in manganese which imparts a greyish or purplish tone to the blue painting, depending on the concentration of pigment. This is most evident in the jewel-like blues of the Jiajing period (1522-1566) which were made by combining local Chinese cobalt with imported cobalt. Unlike copper, cobalt is an ideal pigment for underglaze painting. It is a very strong pigment and does not volatilize even at porcelain temperatures. It is also less affected by oxidation or reduction.[12] It was used not only for blue and white decoration but also in combination with overglaze enamels which will be discussed later in this chapter. Most pigments appear to change colour when they are fired and this is also true of cobalt blue. When it is painted on to the dried but not fired porcelain body, it appears black. It is then covered with a glaze that appears white and seems to obscure the black painting. After firing, the glaze is

---

[12]Hamer, *The Potter's Dictionary of Materials and Techniques*, p. 67.

clear and the painting appears bright blue. All three pigments used in underglaze painting (iron, copper and cobalt) were also used to paint spots either in a glaze or under it. In certain Longquan celadons that were made for the Japanese market, this spotted decoration was very popular and is often referred to by its Japanese name *tobi seiji* or "spotted celadon". At the Jizhou kilns in Jiangxi province, various coloured slips were used for underglaze painted decoration. Here, black, white and brown slips were painted onto a layer of slip which disguised the body and then the vessel was glazed with a colourless glaze. Slip painting was also employed at the Cizhou kilns in north China but usually with just black slip painted on to a white slip ground under a colourless glaze.(figure 37) A later version of slip painted decoration can be seen on some so-called Swatow wares. These have white slip motifs painted on to either a blue or brown ground. The ground on these wares is painted with a coloured glaze, then white slip is painted on top and then the vessel is glazed with a colourless glaze and fired.

Another type of painted decoration used on Chinese ceramics is reserved decoration. This is usually seen on blue and white wares and is really very simple. For painted reserved decoration, the background of a dish will be painted blue for example and areas of the decoration will be left white, or unpainted. The dish is then glazed and fired as usual. The decoration is therefore "reserved" in white. This can be seen on the large blue and white dishes made for the Near Eastern market in the Yuan dynasty (1279-1368) as well as blue and white wares of the Ming and Qing periods. Slip painting can also be used to create reserved decoration. On some blue and white wares, the reserved motifs are painted with slip, then the background is painted in colour around the slip and then the vessel is glazed with a colourless glaze. Reserved decoration can also be produced by the use of a resist. In some Tang *sancai* wares, the pots appear to be decorated with white dots or florettes on a three-colour glaze background.(see figure 13) These were made by making and drying and often biscuit firing the vessel. Then a wax medium was applied to the areas where the decoration is not required. Sancai lead glazes were then applied which did not adhere to the waxed areas. The vessel is then fired and the wax burns off, leaving a resist pattern. This technique is similar to resist dyeing of textiles and the patterns employed resemble textile patterns quite closely. Wax resist is also used to create enamel-on-biscuit decoration which will be discussed below. It should be noted that there are some Chinese ceramics with reserved decoration on a copper red ground. These are rare but can be seen in the imperial porcelains of the Yuan, Hongwu, Yongle, Xuande and some later periods.(see figure 39)

As we have seen, painted decoration is applied to dried or sometimes biscuit fired vessels. Decoration that is incised, carved, impressed or moulded into a vessel is best done when the vessel is still slightly moist so that a smooth line can be achieved. Like painting, *intaglio* decoration is an ancient technique and can be seen on Neolithic as well as Qing period pieces.

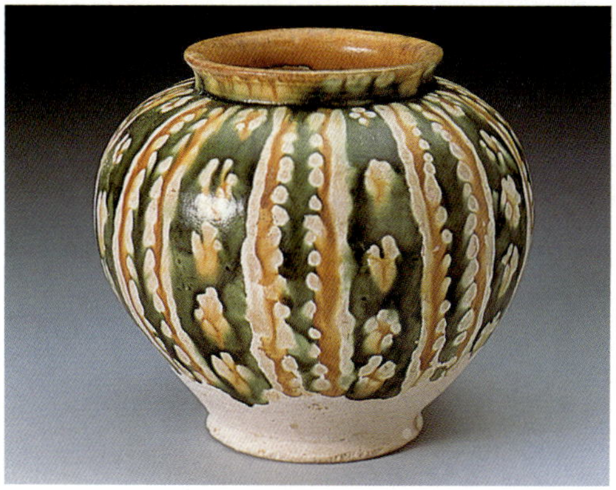

13. Jar with *sancai* glazes, Tang dynasty (618-906). Private Collection.

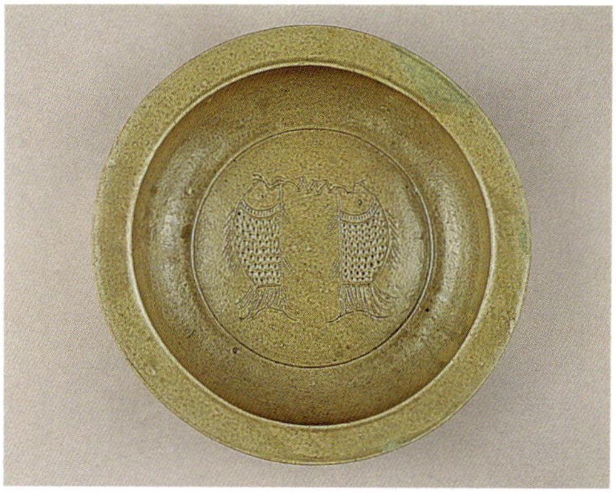

14. Early Yue ware dish with lime glaze, Western Jin dynasty, 3rd-early 4th centuryAD, PDF 250.

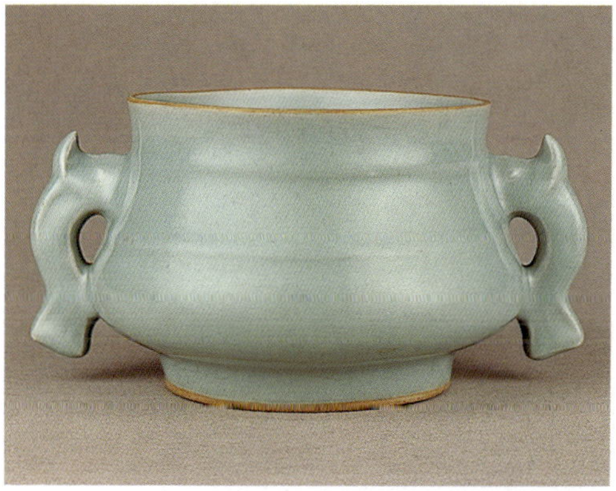

15. Longquan ware censer with lime-alkali glaze, Southern Song dynasty, 12th-13th centuries. PDF 205.

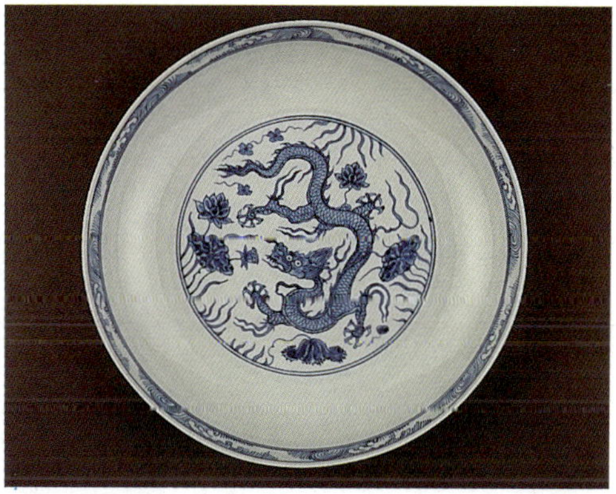

16. Dish with underglaze blue decoration and porcelain glaze, Hongzhi mark and period (1488-1505). PDF A641.

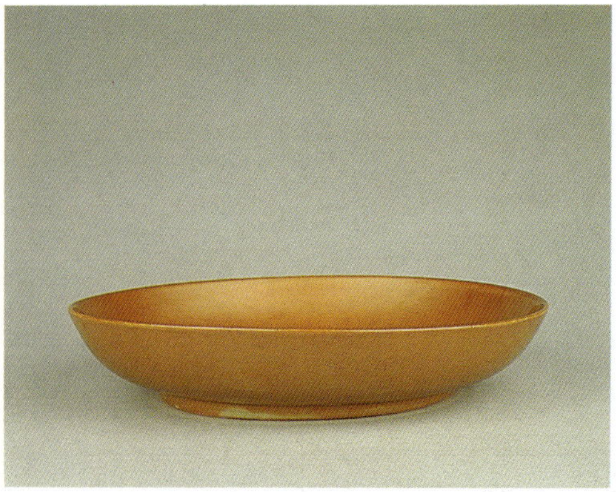

17. Dish with iron brown *café-au-lait* glaze, Ming dynasty, Jiajing mark and period (1522-66), PDF 571.

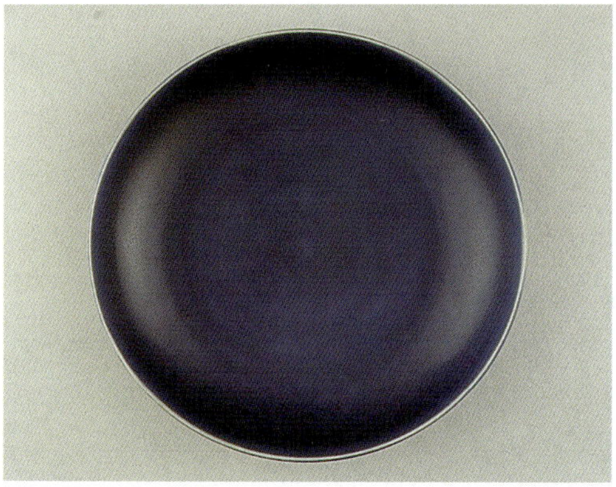

18. Dish with cobalt blue glaze, Ming dynasty, Xuande mark and period. PDF A501.

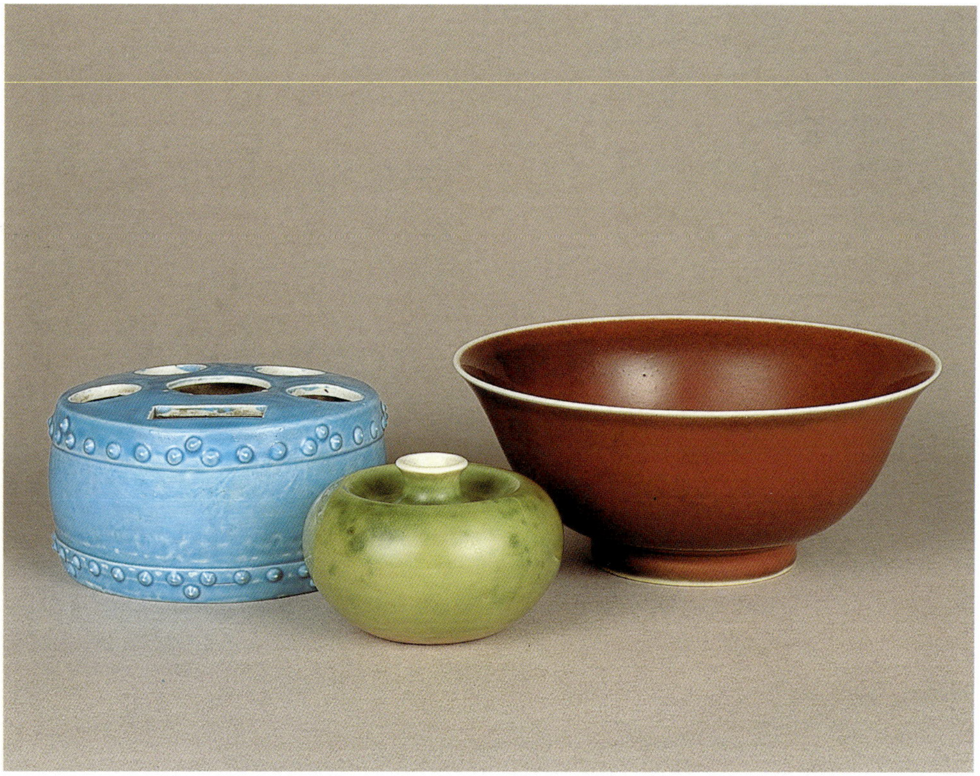

19. Group of three vessels with copper glazes: turquoise ink stick holder, Ming dynasty, early 16th century, green peachbloom water pot, Qing dynasty, Kangxi mark and period (1662-1722), copper red bowl, Ming dynasty, Xuande mark and period (1426-35). PDF A512, A545, A529.

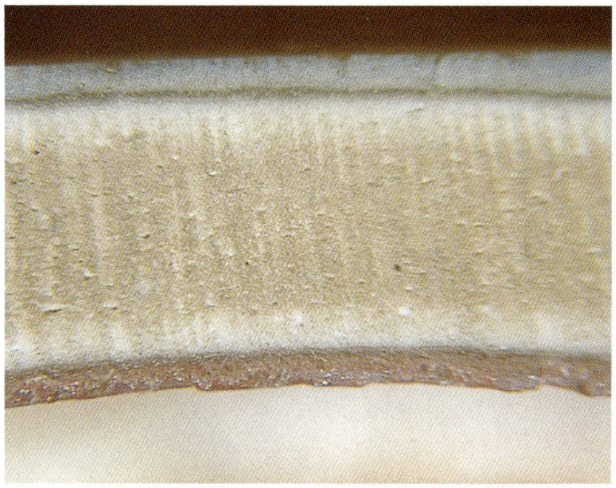

20. Shard of Jun ware vessel showing white layer of anorthite crystals at the glaze-body interface. PDF Study Collection.

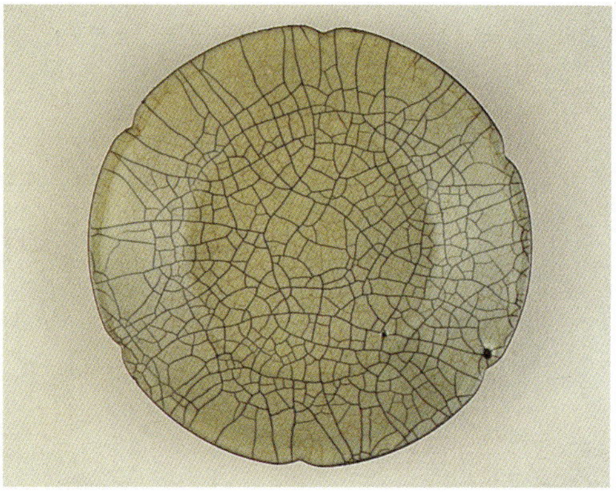

21. Guan ware dish with crackled glaze, Southern Song dynasty, 13th century. PDF 69.

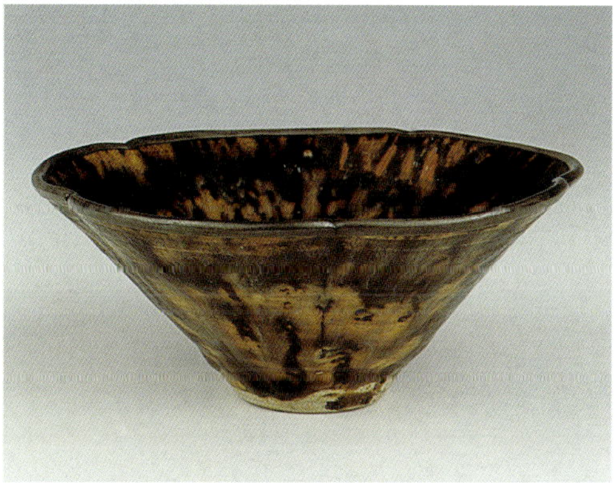

22. Jizhou bowl with tortoise shell glaze, Song dynasty (906-1279). PDF 306.

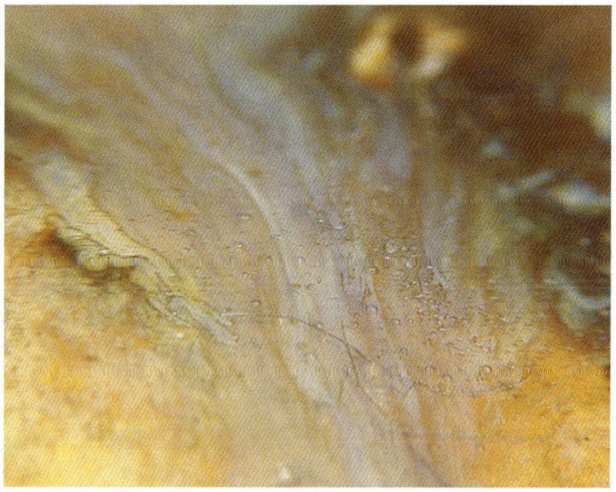

23. Detail of splashes on figure 22 taken through a microscope.

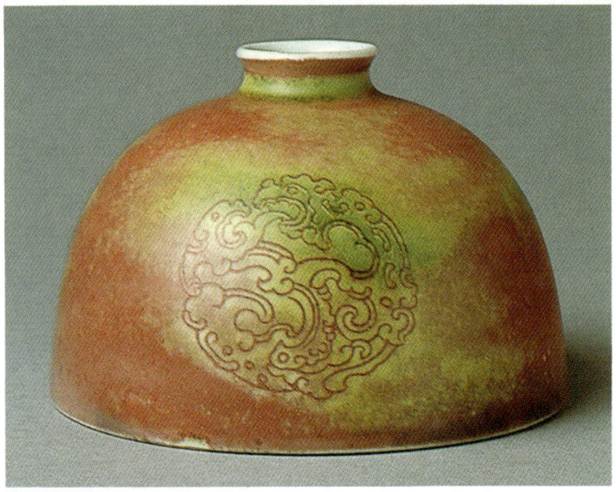

24. Water pot with peachbloom glaze, Qing dynasty, Kangxi mark and period (1662-1722). PDF 580.

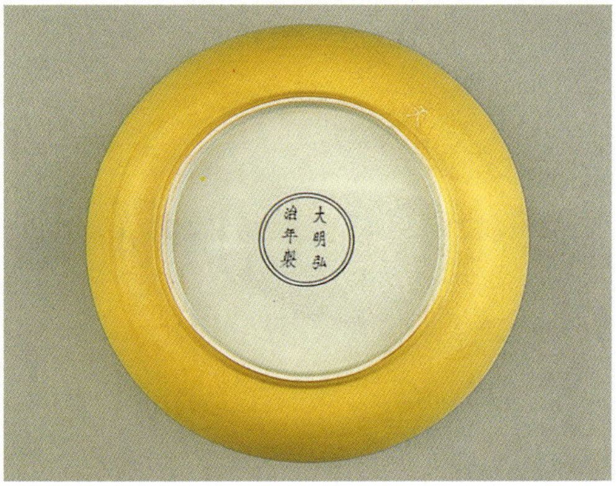

25. Dish with yellow enamel over the glaze, Ming dynasty, Hongzhi mark and period (1488-1505), PDF 599.

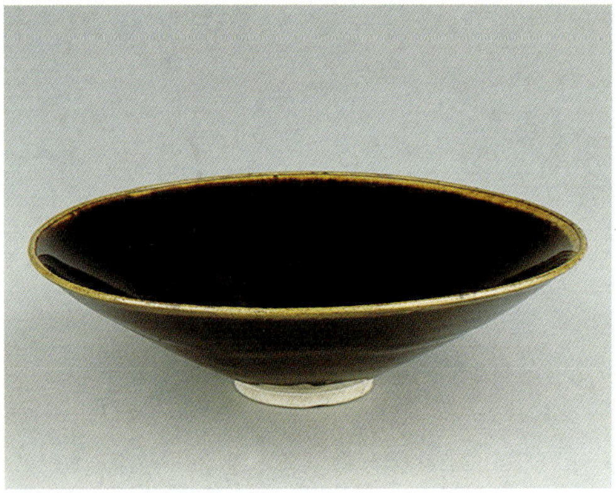

26. Ding ware bowl with black glaze, Northern Song dynasty, late 11th-early 12th centuries. PDF 300.

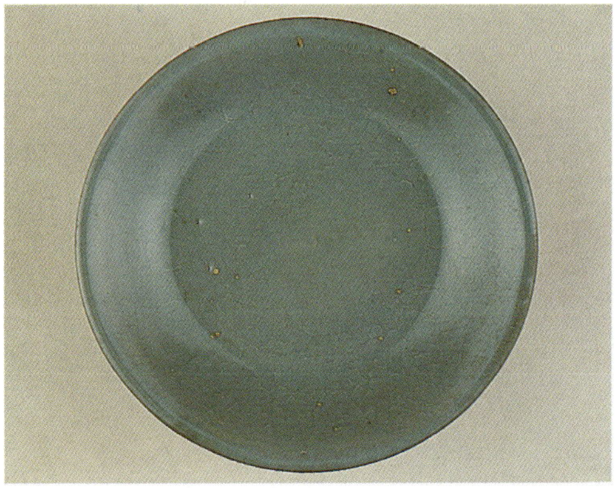

27. Ru ware dish, Northern Song dynasty, early 12th century. PDF 60.

The simplest and easiest form of *intaglio* decoration is that which is impressed into the surface of the clay using various tools such as textile fragments, roulettes or pattern blocks. All of these can be used to decorate a large surface very quickly. Impressing a textile into the surface of a vessel gives it a nice texture. Rouletting can create repeated bands of decoration and pattern blocks can be used to decorate the entire vessel or just one small area. Impressed decoration is most common on pre-Tang ceramics but some early imperial wares have decoration that may have been impressed.(figure 2) A more painterly style of decoration can be achieved by incising or carving decoration into the clay. The incised decoration on Song ceramics is particularly sophisticated. On Ding wares from Henan province, for example, motifs are incised with parallel lines to emphasize the designs and combed to add texture.(figure 28) The parallel lines are used because the glaze on these wares is essentially colourless so it does not enhance the decoration. Deep carving would allow the glaze to pool in the lines but this would not necessarily add colour as the body is also white. The opposite is true of incised decoration on Yaozhou and *qingbai* wares. The decoration on Yaozhou wares is often very deeply carved with an angled tool so the green glaze pools in the crevasses and turns a very dark green.(see figure 29) A nice colour contrast is achieved because the body is also coloured grey. On *qingbai* wares, the body is white but the glaze is a very pale blue which pools in the incisions and turns a deeper blue. Carving can also be used to create pierced decoration.(figure 42) This type of decoration is often referred to as "reticulation". While most commonly found on Ming and Qing imperial porcelains, pierced decoration can also be seen on some Neolithic ceramics. The Jingdezhen porcelain body is suitable for piercing because it can be made quite stiff. Piercing is somewhat risky, however, as it weakens the walls of the vessel thus encouraging collapse. The piercing is therefore usually done when the clay is leather hard.

Carving can also be used to create interesting patterns with slip or inlaid decoration. In Cizhou wares of North China, numerous decorative techniques were used but one of the most innovative was that of "slip carving". With slip carving, a dried vessel is coated with slip and then a pattern is cut through it to reveal the body underneath. The vessel is then glazed with a colourless glaze which tends to make the unslipped areas appear darker. This technique can also be used with two different coloured slips, either black on white or *vice versa*. Usually, the vessel was coated with white slip, then black slip and the black slip was carefully cut away to reveal a pattern against the white slip underneath. A similar type of decoration found on Cizhou wares is inlaid decoration. Here, a design is incised or carved into the body then the vessel is covered with slip. The area with the incised decoration is wiped free of slip leaving only the incisions filled or inlaid with slip. It should be noted that certain glazes can also be carved. On some northern blackwares from Cizhou-type kilns, a thick black glaze has been applied to a vessel and then a design has been cut through the glaze before firing to reveal the unglazed body underneath.

The illusion of carved or incised decoration can be created using moulded decoration. As we have seen, moulds were used in China to create shapes but they were also used to produce decoration when large numbers of similar ceramics were required. With moulded decoration, a design is incised into or cut away from the convex surface of the mould. When the design is incised into the mould, the decoration produced is in relief. When the background is cut away from the design on the mould, the decoration produced on the finished pot is in *intaglio*. The difference between these two can be seen most clearly in a comparison of the moulded decoration of Ding and Yaozhou wares.(see figures 32 and 33) In Ding wares the moulded decoration is usually in relief. In Yaozhou wares, it is usually *intaglio* but the moulds are so deeply carved that the decoration appears to be carved directly into the pot, not moulded. In both instances, the style of moulded decoration can be distinguished from that of non-moulded decoration by the fact that the moulded designs are usually more crowded and stylized. Sometimes both the decoration and the shape are moulded. Some Yuan blue and white dishes that were exported to the Near East for example have elaborate moulded forms and moulded reserved decoration. These would be somewhat easier to paint than the similar dishes with painted reserved decoration as here the reserved areas are actually in relief. Moulded forms and decoration were quite popular and can be seen on Chinese ceramics from the Han to the Qing dynasties.

Moulds can be used to create applied decoration or "appliqués" which belong to the category of relief decoration. Small single moulds for appliqués are called "sprig moulds". To produce an applied motif, a small amount of clay is pressed into a small mould that is the front half of the motif. The back is flat and smoothed out. The appliqué is then released from the mould and the back is then lightly scored and painted with a little slip. This appliqué is then stuck on to the vessel where required to an area that has also been scored and slipped which makes the appliqué adhere better. The vessel is then glazed and fired as usual. Applied decoration can be found on ceramics from the Warring States period (481-221 BC) to the end of the Qing dynasty. Sometimes the applied decoration is left unglazed or in "biscuit". This is most commonly seen on Longquan celadon wares that have moulded red dragons on a green background.(figure 43). Here the dish has been formed, dried and glazed as usual. Meanwhile, a dragon motif has been sprig moulded and laid on top of the unfired glaze. The vessel is then fired. While celadons must be fired in reduction to produce an iron green or blue coloured glaze, in any firing, as it cools down, the atmosphere becomes oxidizing and thus any body material that is unglazed and therefore exposed will oxidize on the surface. The dragon appliqué seen here is actually almost white under a very thin reddish skin. This red "skin" is akin to a layer of rust on a pipe. The red colour occurs because there is some iron in the body material and iron turns reddish or brown in oxidation. While this can occur accidentally on exposed foot rims, where there is red biscuit decoration, the effect is deliberate. It is likely that the red areas were gilded as traces of gold decoration have been found on some surviving examples. Some Longquan wares of the Yuan dynasty have moulded shapes and decoration with panels of biscuit decoration. The panels reveal traces of gilding and were produced by the wax resist method. Gilding has also been found on some Yuan porcelains with white reserved decoration on a blue ground.(figure 34) Some of these were produced by applying a sprig-moulded motif to a blue glazed ground and then glazing again with a colourless glaze so the motif appears white on a blue ground.

Panels of decoration were quite popular in Chinese ceramics, especially in the Jin, Yuan and Ming dynasties (13th-17th centuries). Some panelled wares have rather complicated decoration such as moulded and applied beaded decoration.(figures 40-41) In the Yuan dynasty, many Qingbai wares were decorated with strings of beads that were produced in moulds and then applied very carefully. This beading is sometimes used in conjunction with applied and cut away decoration.(figure 41) Here, the ogival panel is cut out to produce a slightly recessed area into which a moulded or cut out lattice-like floral appliqué is set. Some Ming reticulated bowls have tiny freely modelled figures applied to flat panels.(figure 42)

Relief decoration is not always so complicated. Slip was often used to produce simple but effective decoration in relief. Some of the earliest slip relief decoration is found on ceramics of the late Tang-early Song periods.(figure 36) In these examples, thin slip lines are used to emphasize the shape of the vessel. Later in the Song dynasty, thicker, bolder slip lines are found on northern blackwares. Here light coloured slip lines are trailed on to a black slip ground and then the vessel is coated with a brown glaze. Slip lines can also be used instead of painted or incised decoration on large, vertical vessels where runny glazes need to be contained such as *fahua* ware which is characterized by relief slip decoration which resembles cloisonne metalwork.(figure 38) Here the slip serves another purpose besides decoration in that the coloured glazes used on *fahua* wares are quite fluid and the slip decoration acts as a barrier which helps to contain it. Thin slip lines or patterns were sometimes applied to Yuan blue and white wares or Ming copper reds with reserved decoration to emphasize the white reserved areas.(figure 39) A type of slip decoration that falls into several different decorative categories is that of marbling, which is most often seen on Tang ceramics. This type of decoration can be achieved in three ways: marbled slips, marbled clays or marbled glazes. In the first method, a vessel is coated with a slip of one colour and then another darker one. Before it dries, a tool such as a knife is run along the surface to create a marbled pattern in much the same way that a marbled cake is made. In technique, these are very similar to English slip wares. In the second method, two different coloured clays are combined in the following manner: coils of dark clay and of light clay are rolled together and cut into small, flat pieces about the size of a pound coin. These pieces are then pressed into a mould to form the vessel. After smoothing, a marbled clay vessel is removed from the mould and glazed with a transparent glaze. Sometimes these have a green or amber lead glaze and very rarely *sancai* glazes. Marbled clays can also be combined on the wheel but this technique was used less often in China because the resulting marbling is less distinct than with the moulded method. In the third method, a vessel is coated with a light coloured slip and two coloured glazes are applied one after the other and then marbled with a tool, like the marbled slip method. The glaze method is less satisfactory, however. Marbled decoration continued to be produced on ceramics of the late Song and Yuan periods at the Cizhou kilns. On some Jingdezhen porcelains, especially those with monochrome glazes, *anhua* or "secret decoration" appears which can be incised, slip painted or moulded. This type of decoration is very difficult to see but this is deliberate. It appears most often on early Ming porcelains.

overglaze decoration

The discussion so far has examined decoration that is applied under the glaze. Many Chinese ceramics however have decoration that is applied over the glaze.(see figures 44-50) Decoration is applied over or on a glaze for two main reasons: either the decoration is low firing (i.e. lower firing than the base glaze) or because the base glaze is opaque so any decoration under the glaze would not show through. Most Chinese ceramics with overglaze decoration fall into the first group. The earliest overglaze decoration found on Chinese ceramics thus far is that of Cizhou enamelled wares which were first produced in the late 12th to early 13th centuries. The term "enamel" refers to low firing overglaze decoration. Enamels are very similar to *sancai* lead glazes in that they are mineral pigments in a lead base so they must be fired at a low temperature otherwise they will burn away. *Sancai* glazes, however, are applied directly to the body or slip so they are not "overglaze". Overglaze enamels are usually applied to high fired ceramics. The Cizhou wares with overglaze enamel decoration are high fired stonewares with white slip and colourless glaze. They are fired once for the body and glaze then cooled and painted with enamels. Then they are fired again at a much lower temperature (ca.800°C). The palette for these early enamels is fairly limited and usually consists of an iron red, iron yellow, copper green and an iron-manganese black. The next appearance of overglaze enamel decoration is in the Yuan dynasty on Jingdezhen porcelains. A few shards with overglaze enamels in red and green have been excavated from the Yuan strata at Jingdezhen but they are rare. It is not until the early Ming dynasty (1368-1644) that enamel decoration on porcelain is used more frequently. Recent excavations have uncovered Yongle (1403-1424) period porcelains with overglaze enamel decoration in various colour combinations including: iron red used on its own, red and green and green and yellow. One Yongle period piece combines high-fired iron brown and low-fired green enamel decoration which is the first appearance of high temperature and low temperature colours on one piece. This is a technological precursor to the new enamel palette that emerges in the Xuande period (1426-35), *doucai* decoration.

In the Ming dynasty, there are three different styles of overglaze enamel decoration used: *doucai* and *wucai* which combine overglaze enamel decoration with underglaze blue painting and a third style which is simply overglaze enamels with no underglaze blue, a continuation of the earlier style of enamel decoration, e.g. Swatow wares and those decorated with a single enamel colour.

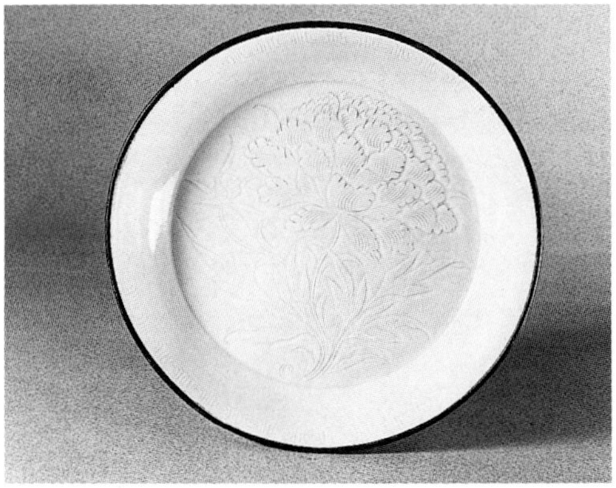

28. Ding ware basin with incised decoration, Northern Song dynasty, 11th-12th centuries. PDF 175.

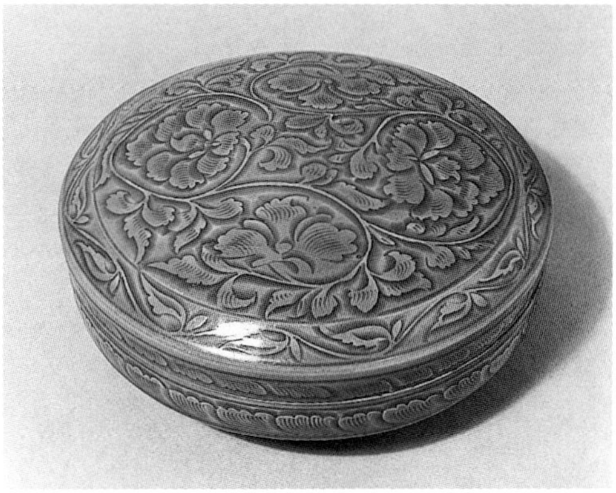

29. Yaozhou ware box with incised and carved decoration, Northern Song dynasty, 11th century. PDF 244.

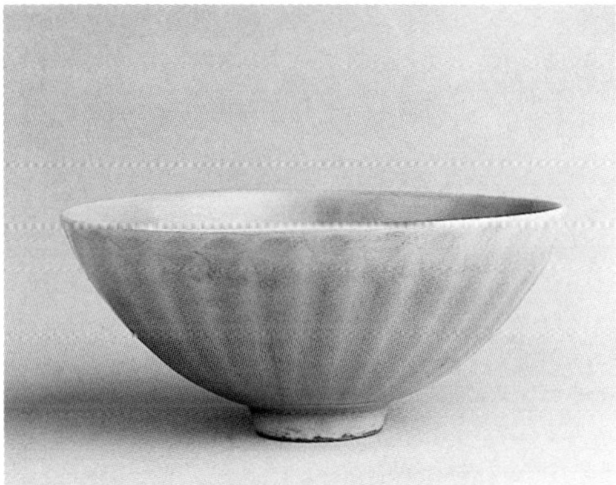

30. Longquan ware bowl with carved petals, Southern Song dynasty, 12th-13th century. PDF 263.

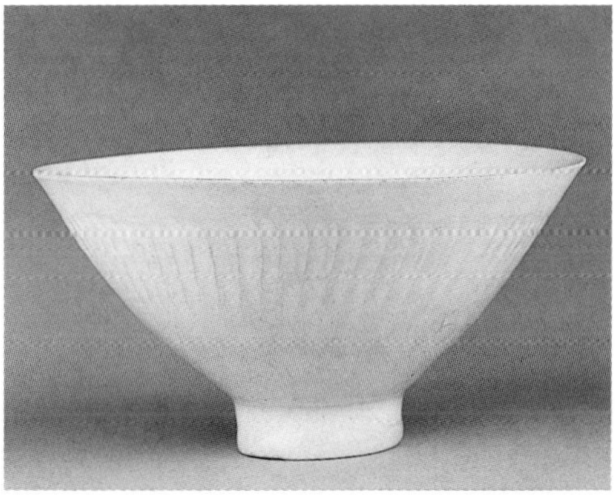

31. *Qingbai* bowl with carved petals, late Northern Song to early Southern Song, 12th century. PDF A459 .

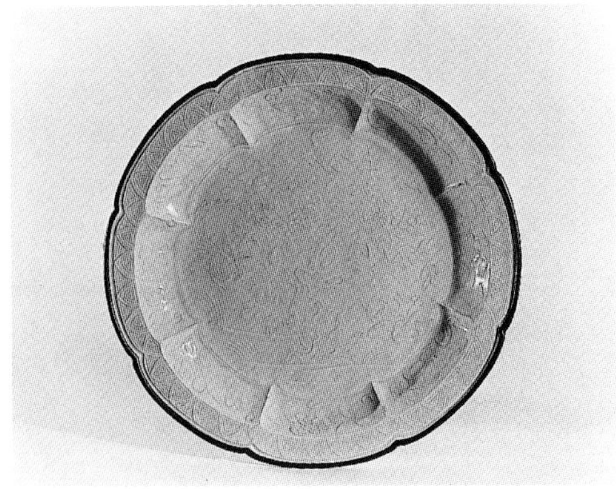

32. Ding ware dish with moulded shape and decoration, Jin dynasty, 13th century. PDF 161.

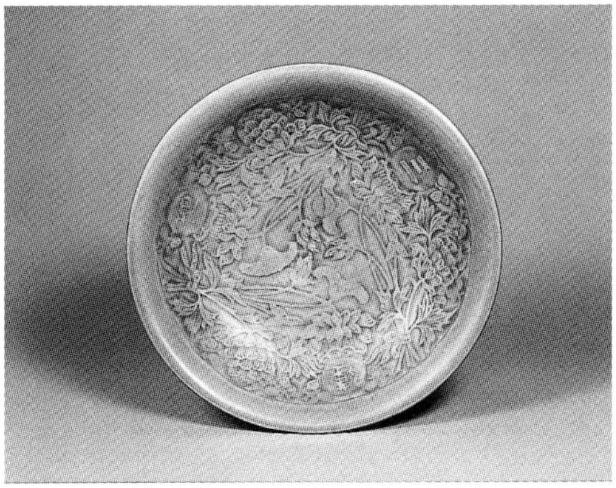

33. Yaozhou ware bowl with moulded decoration, Northern Song dynasty, 12th century. PDF 245.

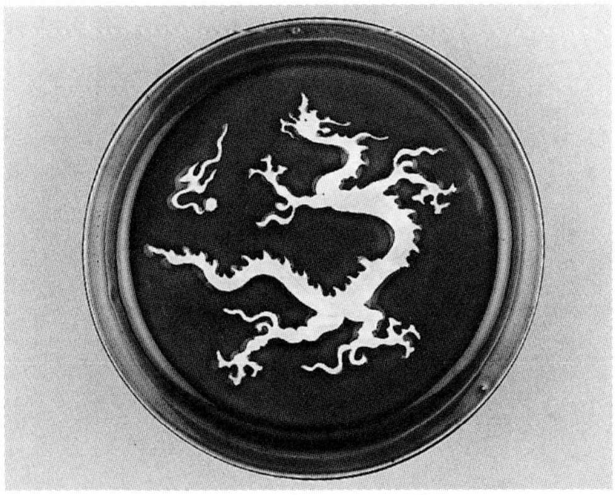

34. Jingdezhen porcelain dish with applied decoration, Yuan dynasty, mid-14th century. PDF A562.

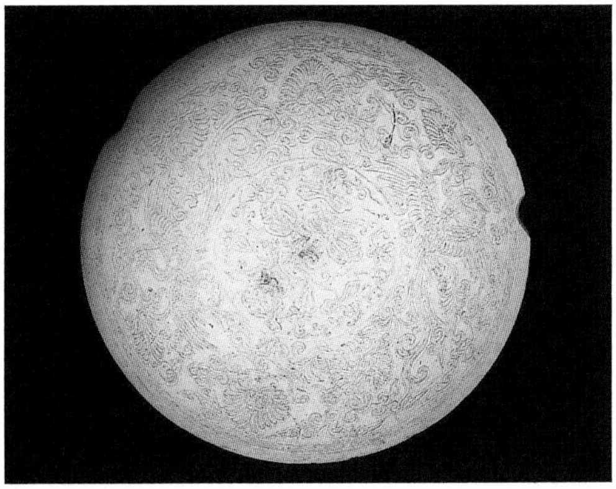

35. Mould for Ding ware bowl, dated by inscription to AD 1184. PDF 181.

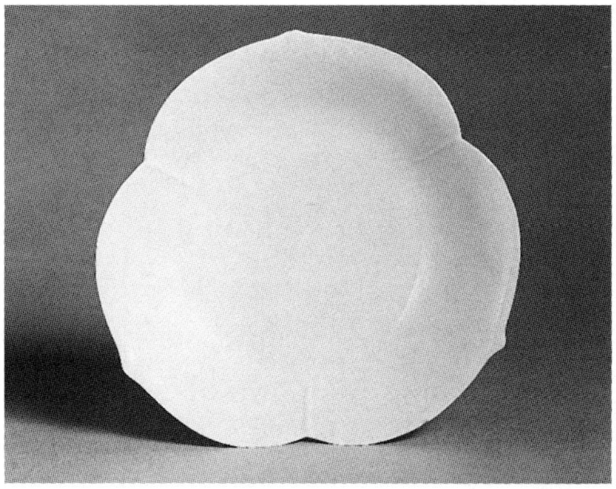

36. Ding ware dish with slip trailed decoration, Northern Song dynasty, 10th century. PDF 173.

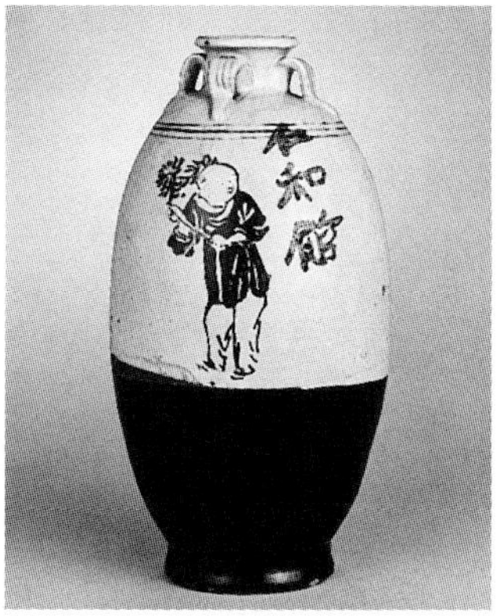

37. Cizhou-type jar with slip painted decoration, Yuan dynasty (1279-1368). PDF 316.

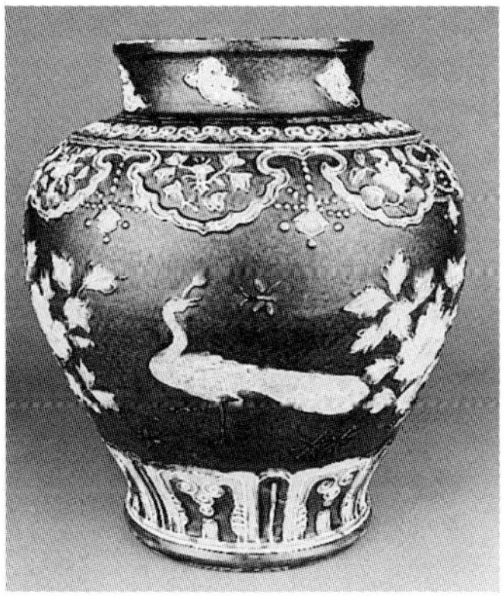

38. *Fahua* jar with slip decoration, Ming dynasty, Hongzhi period (1488-1505). PDF 759.

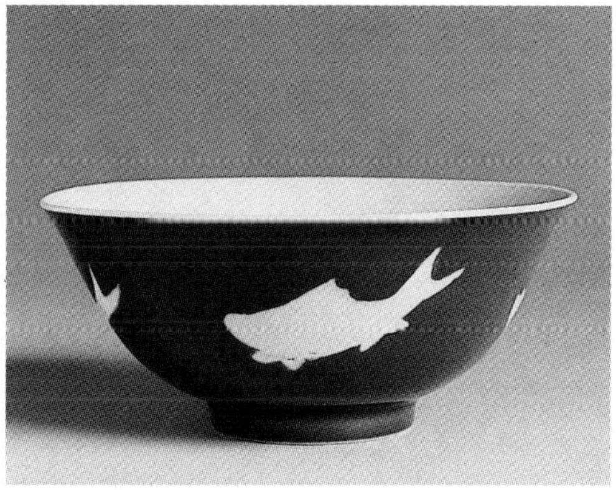

39. Silhouette ware bowl with slip decoration, Ming dynasty, Zhengde mark and period, (1506-21). PDF A526.

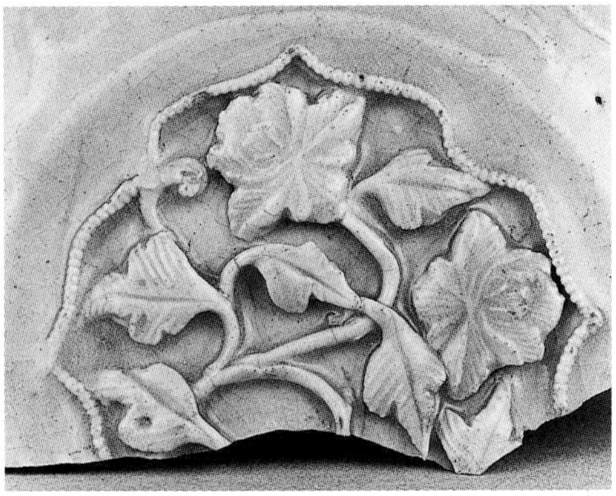

40. Detail of a shard of *qingbai* ware with applied decoration. PDF Study Collection.

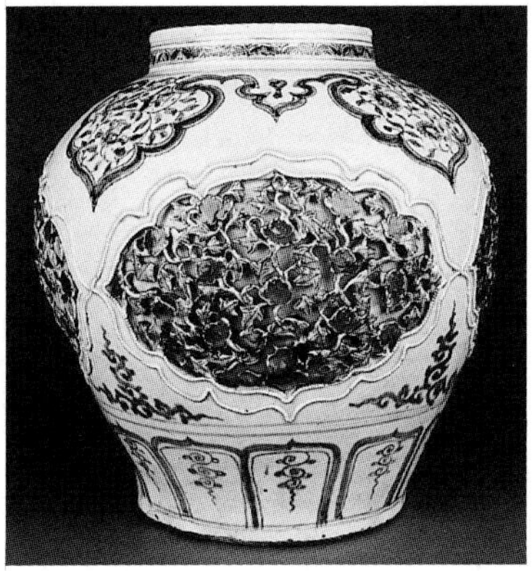

41. Jingdezhen porcelain jar with underglaze blue and red painting and applied decoration, Yuan dynasty, early 14th century. PDF B661.

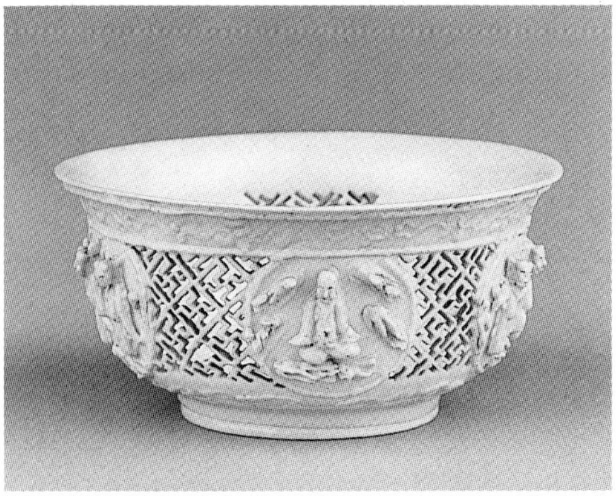

42. Porcelain bowl with reticulation and applied biscuit relief decoration, Ming dynasty, Wanli period (1573-1619). PDF A403.

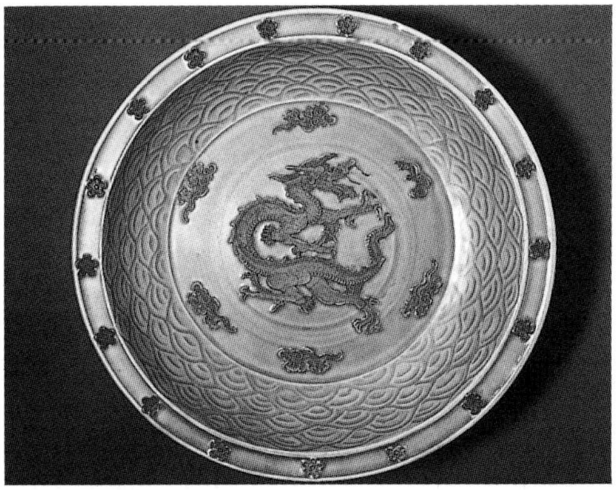

43. Longquan ware dish with sprig-moulded biscuit decoration, Yuan dynasty, 14th century. PDF 255.

The development of the *doucai* style of decoration is associated with Chenghua period porcelains (1465-87) but in fact it made its first appearance in the Xuande period (1423-35). The *doucai* style combines both high fired underglaze blue decoration and low fired enamel decoration in one piece.(figure 44) To make a *doucai* style dish, one would first form the dish on a wheel then let it dry before painting in the outlines of a design in underglaze cobalt blue. The dish would then be glazed and fired at a high temperature. After cooling, overglaze enamels would be painted into the outlines of the motif on the fired dish. The dish would then be fired a second time at a low temperature. The enamels are thus "juxtaposed colours" which is how *doucai* is often defined. The palette is similar to that of Cizhou enamels but with a few more colours including aubergine and brown.

After the Chenghua period, a slightly different style of enamel decoration was developed which was somewhat easier to produce: *wucai* or five-colour decoration. In this style of enamel decoration, some aspects of the design are painted in underglaze blue but most of it is painted in overglaze enamel.(figure 45) Because most of the design is painted on top of the fired glaze, if a mistake is made in the decoration, most of it can be wiped off without ruining the pot. Underglaze blue painting is absorbed by the clay body so it cannot be wiped off. Ideally an overglaze blue enamel would have been the best option but this was not developed until the Kangxi period (1662-1722) (see discussion below). Both the *doucai* and the *wucai* styles of enamel decoration were also used in the Qing dynasty (1644-1911) but the development of new enamel colours in the 18th century enabled potters to produce much more colourful enamelled wares at this time, including pastels and pinks which were not possible in the Ming period.

This new enamel palette, which was developed in the early Qing dynasty, is perhaps more familiar in the West than the earlier Ming styles. Both the *famille verte* and *famille rose* styles of enamel decoration were developed in the late 17th and early 18th centuries.(see figures 46, 48) The *famille verte* palette is characterized by the use of transparent enamels. The Chinese name for this palette is *yingcai* (hard colours). The colours include: green, yellow, purple, turquoise, brown, red and black. It is the predominance of green enamel that gives this palette its name. Sometimes gold enamel is also included which is the first time fired-on gold has been used in Chinese ceramics.(figure 47) Previously, when gold was applied to ceramics such as Song Ding wares or *kinrande* wares of the 16th century, it was gold leaf affixed with an adhesive. This method was not very satisfactory as the gold did not adhere well so very often all we are left with is the shadow of the gold design in the remaining adhesive. In the late 17th century, however, a gold enamel was developed that could be fired on thus producing more permanent gilding. The *famille verte* palette also includes some underglaze blue decoration, much in the same way as the *wucai* style of decoration.(figure 46) However, at some time in the Kangxi period (1662-1722), an overglaze blue enamel was developed that obviated the use of underglaze blue and enabled potters to create porcelains with an entirely overglaze design.(see figure 47)

The development of overglaze cobalt blue enamel coincided with the introduction of a new body material and glaze at Jingdezhen. More kaolin was added to the body so that the ratio of kaolin to china stone was about 50:50. As we have seen in chapter one, this is an ideal proportion because the body is then both plastic and workable. By adding more kaolin to the body, the firing temperature was increased because adding more clay raises the amount of alumina in the body which fires at a higher temperature than silica. As the body was now higher firing, the glaze had to be adjusted so the amount of lime or calcia flux had to be reduced. By reducing the lime content, the proportion of alkali fluxes accordingly increased thus rendering the glaze a classic lime-alkali glaze. As we have seen in chapter two, lime-alkali glazes tend to be somewhat opaque and this new porcelain glaze was no exception. An opaque white glaze was highly desirable in this case because overglaze enamels look best against a very white ground. However, underglaze blue decoration does not show through an opaque white glaze so the development of an overglaze blue enamel was imperative. There are a few instances of the use of a low temperature cobalt blue glaze or enamel in the Ming dynasty but technical problems ensured that its use was infrequent. The high manganese content of local Chinese cobalt creates a purplish blue in a lead glaze (which is what enamels are). This can be overcome by increasing the alkali content of the base glaze but to ensure that this glaze melts at the same temperature as the other enamels, its lead content had to be decreased accordingly. This problem does not occur with the high iron imported cobalt which was used in earlier periods. This new blue enamel recipe may have been adopted from cloisonné metalwork as the enamels used there were being produced for the court in the imperial glass workshops[13] which also provided enamels for porcelains.

As we have seen, in the early 18th century, the overglaze enamel palette changed dramatically with the introduction of new opaque enamels and a also a pink enamel which gave rise to the term *famille rose*.(figure 48) The new opaque colours were made from an opaque white which was made from lead and arsenic. This could be mixed with other enamel colours to produce a range of pastels that were not possible before. This palette also include an opaque yellow (lead and tin, not antimony--figure 49) and a transparent rose pink that was made from crushed ruby glass in a clear enamel base. This rose pink was mixed with white to produce a pale pink enamel.(figure 50) The composition of these new enamels is close to that of the new overglaze

---

[13]Scott, R., "18th century overglaze enamels: the influence of technological development on painting style" in *Style in the East Asian Tradition, Colloquies on Art and Archaeology in Asia No. 14*, Percival David Foundation, 1987, p. 155-56.

blue enamel and thus may also have been adopted from cloisonné enamels. The Chinese name for this palette is *fencai* (powder colours) or sometimes *yangcai* (foreign colours).

Both the *famille verte* and the *famille rose* palettes are primarily overglaze enamels but another type of enamel decoration was used in both the Ming and Qing dynasties: "enamel-on-biscuit". This style of decoration consists of overglaze enamels painted on to a fired but unglazed body.(figure 61) The technology is very similar to that of Han and Tang lead glazed wares however they were applied to fired but unglazed bodies and enamels are essentially low firing lead glazes. Like the earlier lead-glazed wares, the Ming and Qing wares with enamels on the biscuit have a very different appearance than those with overglaze enamels. Enamels applied directly to the biscuit are somewhat darker than overglaze enamels. They are also less shiny. This type of decoration was used in several ways: as a monochrome glaze, in combination with other enamel colours or as part of the decoration of a dish with a high-fired glaze.(see figure 61 above) In the latter type, the biscuit area is reserved by use of wax resist. A design is carved into the leather hard but not fired body. Wax is then applied to the design and the dish is then glazed and fired. The wax burns off in the firing leaving a reserved, biscuit design. This design is then painted with a low temperature enamel and then fired again at a low temperature in a muffle or enamel kiln. This type of enamel decoration is most commonly seen in Ming ceramics.

in-glaze decoration

There are some decorative techniques which do not fall into any of the above categories because they are applied neither under nor over the glaze. These can be grouped together as "in-glaze" decoration in that they are usually applied to the unfired glaze. Examples of in-glaze decoration include splashed decoration (see cover illustration) which can be seen on Tang "Jun" and Song and Yuan Jun ware, Jizhou wares with tortoise shell or paper-cut decoration and Ming and Qing "silhouette" wares.(see figures 22, 39) The so-called Tang "Juns" were made at kilns in Henan province which began to produce Jun wares in the Song dynasty.(see map) The Tang "Jun" wares are classified as blackwares because they are characterized by black glazes with "splashes" of a lighter ash glaze. Splashed decoration is quite simple in that wood ash is "splashed" or brushed on to an unfired base glaze which when fired, causes these areas to change colour. With Jun wares, the Tang examples have milky blue or white splashes on a black base glaze and latter Jun wares have purple copper splashes on a blue Jun glaze background. Usually, the copper is applied randomly however there are some examples of Jun wares with deliberate designs in copper that appear to be carefully painted on.

A range of in-glaze decorative techniques were used at the Jizhou kilns in Jiangxi province. Jizhou "tortoise shell" glazes have light dots or splashes on a dark glaze background. Both the splashes on Tang "Jun" wares and Jizhou tortoise shell glazes have a tendency to turn bluish. This is due to the action of ash on the base glaze which causes areas of localized phase separation.(figures 22, 23. See chapter two for a full explanation of phase separation.) With paper-cut decoration, paper stencils are placed on to a dark glaze background and then a light coloured ash glaze is splashed or flicked on top with a brush. The stencils are then allowed to burn away in the firing leaving dark designs on a light yellowish background. These wares are fired in reduction. Also at Jizhou, light slip colours were painted on to a dark glaze background in patterns that resemble *guri* lacquer. Sometimes resist decoration was made with paper stencils at Jizhou whereby paper designs were affixed to the dried vessel which was then dipped in glaze. Before firing the paper was removed so the finished vessel has biscuit reserved motifs against a dark glazed background. These areas were sometimes enhanced with brown iron painting. The most interesting in-glaze decoration used at Jizhou is that of "leaf temmoku" which is made by affixing a leaf to the inside of a dried bowl then dipping it in a dark iron glaze and then firing. In the firing the leaf burns away and leaves a pale shadow leaf, like an x-ray. These wares are fired in oxidation so the carbon in the leaf completely burns away. This is the opposite effect of paper-cut decoration which is fired in reduction so the carbon is not burned away and thus leaves a dark pattern instead of a pale one.

In the Ming dynasty, another type of paper resist decoration was produced on imperial porcelains that is not completely understood, even today. These are porcelains with so-called "silhouette" decoration of either white on red or red on white.(figure 39) The 16th century red on white examples were probably made in the following manner: the vessel was shaped and dried; it was then dipped in colourless porcelain glaze, dried and a paper stencil was affixed to the surface. A copper red glaze was then painted onto the revealed area and the vessel was fired. It is also possible that the red glaze was sandwiched between two clear glazes or pigment was applied under the clear glaze.[14] Silhouette wares were made originally in the Xuande (1426-35) and Zhengde 1506-1521) periods but like other Ming styles were copied in the Qing dynasty.

[14]Wood, N., *Chinese Copper Red Wares*, PDF Monograph Series No. 3, p. 26.

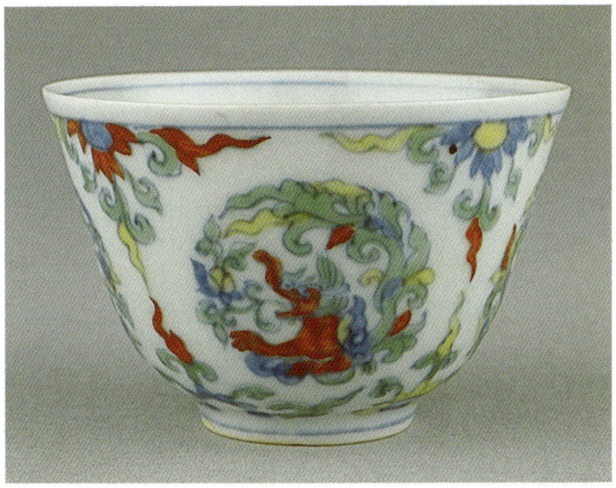

44. Wine cup with *doucai* enamel decoration, Ming dynasty, Chenghua mark and period (1465-87). PDF A745.

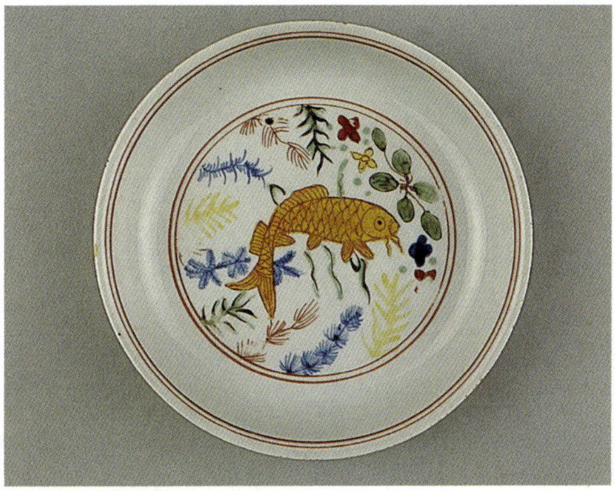

45. Dish with *wucai* enamel decoration, Ming dynasty, Jiajing mark and period (1522-66). PDF A731.

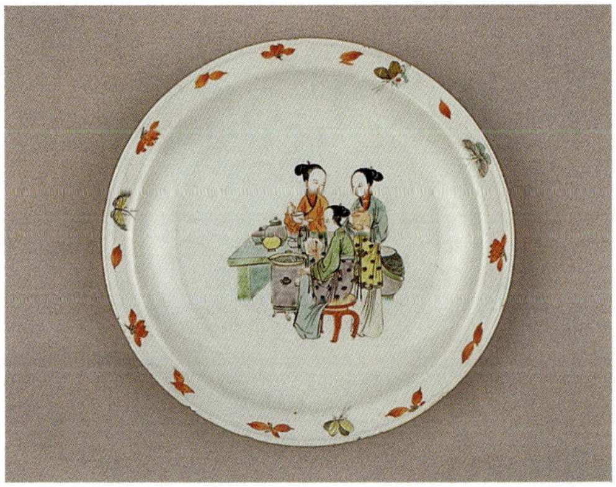

46. Dish with *famille verte* enamel decoration, Qing dynasty, Kangxi period, 1st quarter of 18th century. PDF 805.

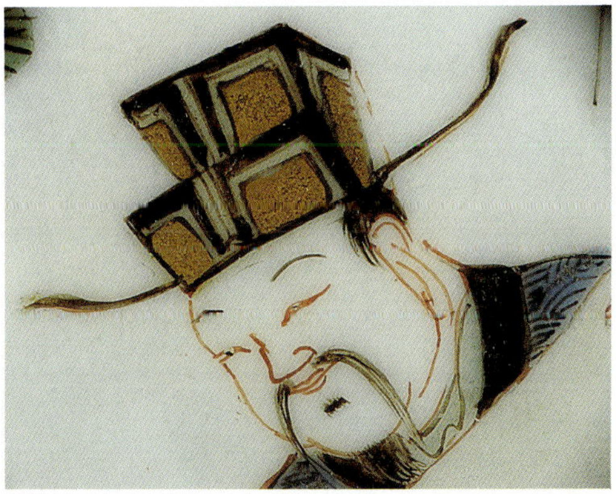

47. Detail of an early 18th century dish with overglaze enamel decoration showing gold enamel and overglaze cobalt blue enamel. PDF A821.

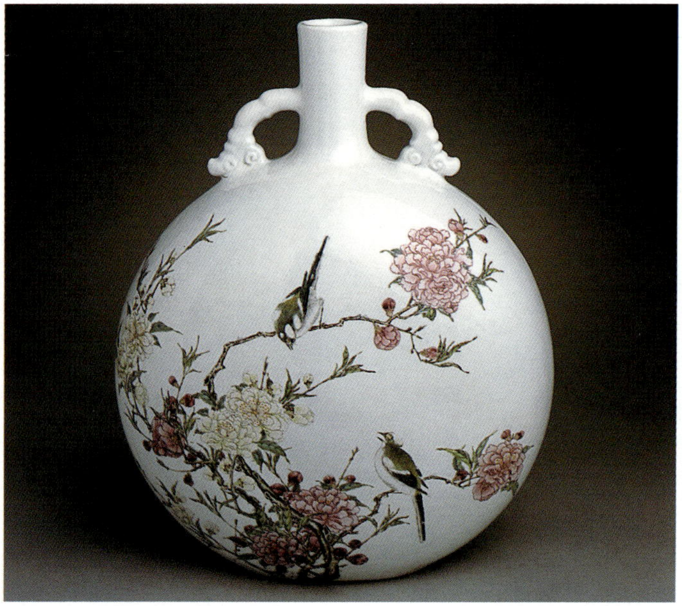

48. Flask with *famille rose* enamel decoration, Qing dynasty, Yongzheng mark and period (1723-35). PDF 824.

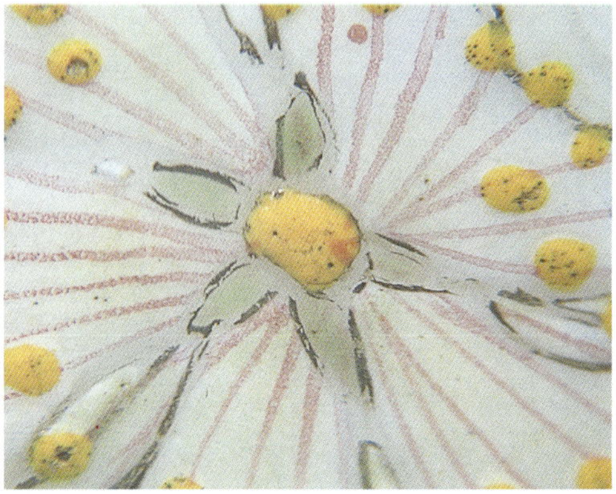

49. Detail of figure 48 showing yellow and white enamels.

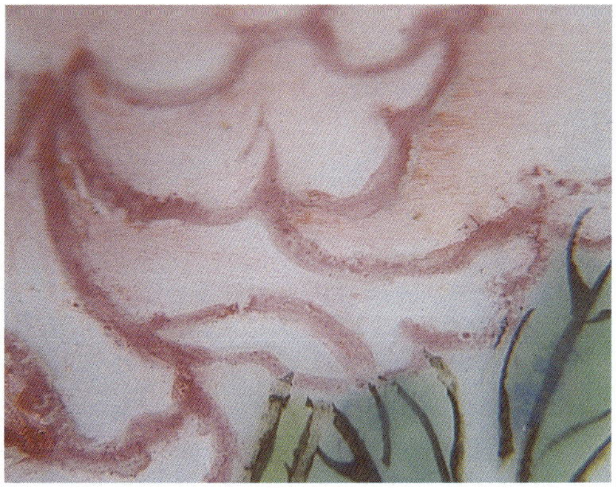

50. Detail of figure 48 showing rose pink enamel.

4. Kilns and Kiln Furniture

As we have seen, ceramics are made from clay and water. In order to transform clay into a hard, ceramic material, this water must be removed. The first step is drying. The second step is firing or baking in a kiln. A kiln, therefore, is like an oven and its function is very much the same. In fact the word "kiln" is derived from the Latin for kitchen, *culina*. Like an oven, a kiln must be able to heat up the pots and then hold a high temperature for a long enough period to remove all the water from the clay and then force the clay particles to move closer together and fuse. The effects of firing will be discussed fully in chapter five. Like a fire in a fireplace, a kiln requires air to burn. As the fuel burns, it creates a "draught" or a current of air. A chimney is required to create the draught and in a fireplace, move smoke upwards and out of the house. Kilns also have chimneys and draughts and are often classified by the direction of the draught, usually up, down or across the firing chamber where the pots are placed.

The earliest method of firing pots used by man was not in kilns at all but rather bonfires. Pots were placed on the fire and covered by wood or grass or other fuel. As may be imagined, this type of "kiln" is quite primitive and not very efficient. A bonfire cannot reach very high temperatures because it is exposed to the elements which also means that a constant high temperature cannot be achieved. The heat is not contained and much of it is lost. Ceramics fired by this method tend not to be very hard and they are often blackened where exposed to direct heat and flames. It is not suitable for firing glazed ceramics, therefore. Ceramics fired in bonfires are quite functional, however. They are classified as earthenwares because they are fired at a low temperature and therefore quite porous but they can be used for cooking or storing food and liquids. This method of firing is still used today in some areas of the world because it is cheap and relatively easy.

The first real kiln that was developed is slightly more efficient and sophisticated than an open, bonfire firing. This type is known as a "pit kiln". Pit kilns are dug into the ground and fuel is placed beneath and on top of the pots in the pit. These kilns are more efficient than a bonfire because the heat is partially contained and thus a higher and more constant temperature can be achieved. The pots are still exposed to the fire however. A better kiln design would enable the heat to be contained and would protect the pots from the fire. This is the earliest type of kiln used in China and is known as a *Yangshao* kiln, after the early neolithic culture that used it. This is a kind of "updraught" kiln and is basically a tunnel dug into the ground with a cylinder on top. This is similar to what is known as a "bank kiln" or a tunnel dug into the side of a hill, but smaller. Here the fuel is placed at the bottom in the "fire box" or area which contains the fuel and the heat is drawn upwards to the other end by a chimney. The fuel in the fire box is also separated from the pots in the firing chamber. Updraught kilns are therefore quite efficient. They heat up slowly and reach fairly high temperatures (1000°C) while the heat is more evenly distributed throughout the firing chamber than in a pit firing. The firing is more controlled and produces better pots as a result.

A development from the *Yangshao* updraught kiln is the "cross draught" kiln. In north China, these were known as *mantou* kilns because they are shaped like a domed bread bun.[15] The *mantou* kiln is a very efficient kiln design and was used all over north China from the Shang dynasty (c.1700-1100 BC) onwards. Like the pit kiln, this is partly dug into the ground but most of the kiln is usually above ground. The firing chamber is domed and horse-shoe shaped in section and is situated between the fire box and the chimney. Thus the draught moves the heat across the kiln and the pots are separated from the fire by a "bag wall". The single cylindrical firing chamber is usually made of mud and bricks and is very thick and very well insulated because it is lined with a high firing or "refractory" clay. Though its capacity is limited because it only has one firing chamber, it is very efficient. It takes a long time to heat up and cool down, can reach very high temperatures (1350°C) and uses wood or coal for fuel. This type of kiln is responsible for some of the most spectacular Chinese wares including Ru and Jun wares.

As we have seen, the clay in the north of China is quite different from that in the south. The same is generally true of kiln design. The *mantou* kiln is a northern kiln design. In the south, long bank or tunnel kilns were preferred. Because they typically snake up the side of a hill, these kilns are known as "dragon kilns" in China. The kiln is essentially a very long chamber with a fire box at one end and stoke holes along the sides up to the other end so the draught moves up through the kiln, as the kiln itself slopes upwards which directs the draught instead of a chimney. The characteristics of this type of kiln are: large capacity (much larger than *mantou* kilns) and rapid firings. The capacity of these kilns is huge because they are often 60-100 metres long. Sometimes they are divided into separate chambers ("multi-chambered") or stacks of saggars or firing boxes are used to separate different groups of wares. The stoke holes running along the side of the tunnel enable the fire to be continually stoked and they therefore heat up very fast. Because the firing is started at the bottom and stoked along the way, after each stoking the previous chamber begins to cool down and thus these kilns also cool rapidly. Of course, this has an effect on the type of ceramics produced and thus dragon kilns are most suitable for wares that require relatively rapid firings, like some celadons.(see figure 15) In dragon kilns and most other kiln types, the ceramics are raised off the floor of the firing chamber so that the draught can circulated under and around the saggars.

---

[15]The three main kiln types used in China are very well illustrated in several of the books listed in the bibliography including: *Iron in the Fire* from the Oriental Ceramics Society and *Ceramic Masterpieces* by Kingery and Vandiver.

The disadvantage of a continually stoked and cooled firing is the production of a variable atmosphere. There are three types of kiln atmosphere: reducing, neutral and oxidizing. The atmosphere refers to whether there is a lot of oxygen present, some oxygen, or very little. When we light a fire in a fireplace, we can produce a nice clean fire or a smoky one. A fire gets too smoky when there is insufficient air circulation (when the flue is closed for example). A similar situation occurs in a kiln. If there is a lot of air, and thus oxygen present in a kiln, the atmosphere is described as "oxidizing". The opposite is a smoky atmosphere which is called "reducing". The atmosphere in the kiln has an effect on the colour and hardness of ceramics so it needs to be controlled. As we have seen, celadons require a reducing atmosphere whereas Ding wares, which have a warm toned glaze, require an oxidizing atmosphere.(see chapter five for full discussion) In a dragon kiln, fuel is constantly being fired or burned and cooled down so the atmosphere varies throughout the firing period. The pots created are therefore less consistent in quality than those produced in a *mantou* kiln. Another disadvantage of dragon kilns is that they are wood-fuelled. Wood is an expensive fuel and tons of it is required to fire up such a huge kiln. In China, huge areas were deforested because of these kilns. However, it should be noted that the beautiful, unctuous Longquan celadons could only have been produced in wood-fuelled dragon kilns.

Perhaps the best kiln design used in China was that of the "egg-shaped" kiln used at Jingdezhen from about the end of the 16th century.(figure 51) This kiln combines the best features of both *mantou* and dragon kilns in one kiln. It has a very large, egg-shaped single firing chamber, up to about 10 metres in length and a very tall chimney. The fire box is located at the wide end of the chamber, opposite the chimney.[16] The large chimney is important because it creates a very strong draught and the kiln therefore heats up rapidly to very high temperatures. The firing schedule is therfore quite short, about 36 hours.[17] These kilns are also very well insulated. There are vents along the top part of the kiln which are used to control the atmosphere, not to stoke the fire as with dragon kilns. In fact, these egg-shaped kilns require much less wood fuel than dragon kilns and are therefore cheaper to run. The shape of the kiln, with a wide, tall, domed front section and a narrower, lower opposite end allows for areas of different temperature in one kiln chamber. The cool zone was near the chimney at the back and the hot zone was near the fire box at the front. This means that different types of wares can be fired in the same firing. For example, high firing blue and white porcelains would be placed in the hotter zone and lower firing turquoise monochromes in the cooler zone. This was particularly important in the late Ming and Qing periods at Jingdezhen when a very wide range of imperial ceramics were being produced for the court.

---

[16]Vainker, p. 222.

[17]Kerr, *Chinese Ceramics: Porcelain of the Qing Dynasty*, p. 39.

In all types of kilns, the ceramics are not just placed directly on the floor of the chamber. They are usually raised up on a layer of saggars which itself is placed on a bed of sand. In single chambered kilns, walls of saggars were used to divide the ceramics and direct the flow of the draught and heat. A saggar is a clay box which is made of a highly refractory clay and is useful for protecting the pots from heat and dirt. Within the saggars, the objects were again placed on kiln furniture such as rings or discs of clay on a bed of sand or grit. The marks of kiln furniture can be seen on some ceramics. On Ru wares, for example, there are usually "spur marks" on the bases of vessels.(figure 53) These are little rice-grain or sesame-seed shaped marks which are left on the glazed base by a crown-like ring of clay called a setter. This setter has little spikes or "spurs" on it which the vessel rests on. (figure 53) In this way, the pot is raised up off the bottom of the saggar and the base can therefore be glazed. Pots can also be fired on their foot rings or on their mouth rims as can be seen with Ding wares. Several bowls can be fired in one saggar if turned upside down and fired on their mouthrims (figure 52) on "steps" that line the walls of the saggar. These stepped saggars are quite handy as they allow a number of bowls to be placed in one saggar and thus increase production. However, because these bowls are stacked, the feet have to be very small to allow for space and the mouth rims must be wiped free of glaze. The resulting unglazed mouthrims were usually bound with a copper alloy after cooling. Another advantage of these types of saggars is that Ding wares are made very thin and thus are prone to collapse if fired on their feet. The wider mouth rim can take more weight. This method of firing is known as *fushao* in Chinese and was also used for firing *qingbai* wares. Kilns and kiln furniture are therefore extremely adaptable and it is clear that the Chinese potter was able to use his empirical knowledge to create ever more efficient kiln designs throughout history. An egg-shaped kiln is in fact still used today at Jingdezhen.

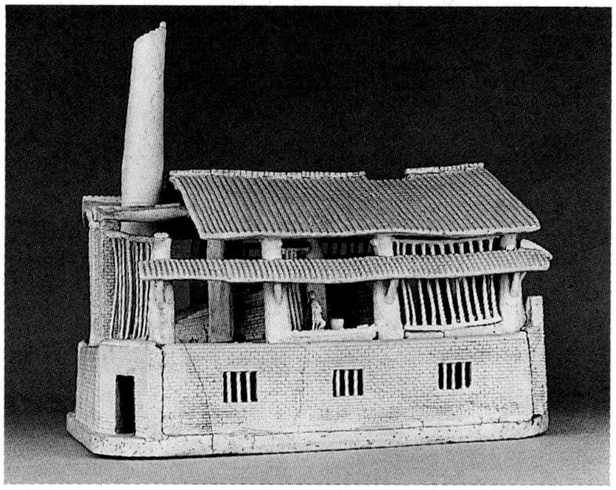

51. Model of a Jingdezhen egg-shaped kiln. Private Collection.

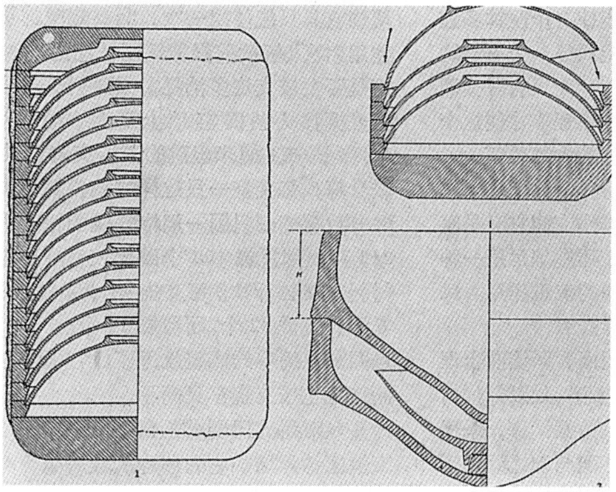

52. Stepped saggars of the type used at the Ding kilns and for *qingbai* ware.

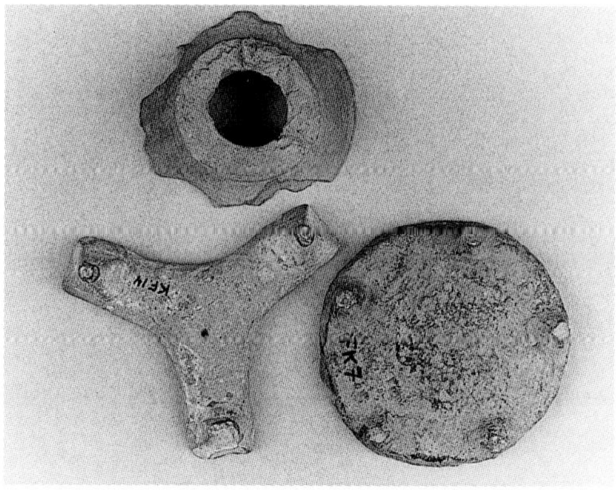

53. Kiln furniture. PDF Study Collection.

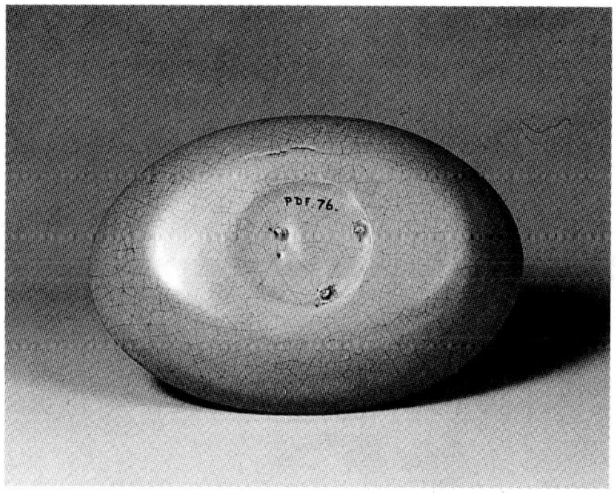

54. Spur marks on the base of a Ru ware dish. PDF 76.

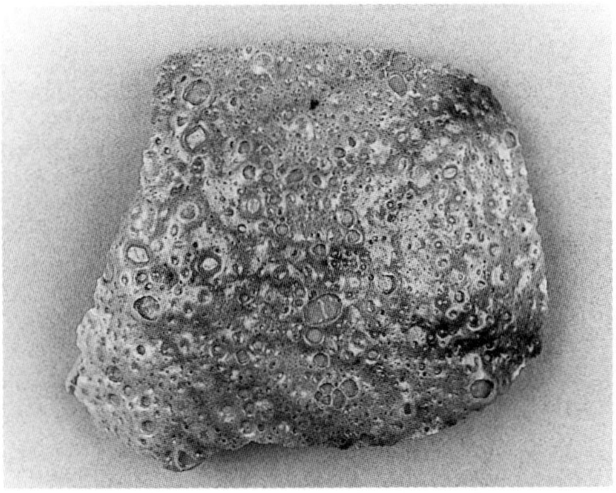

55. Shard of a dish with blistered glaze. PDF Study Collection. Blistering occurs when bubbles in a glaze do not smooth out in the firing.

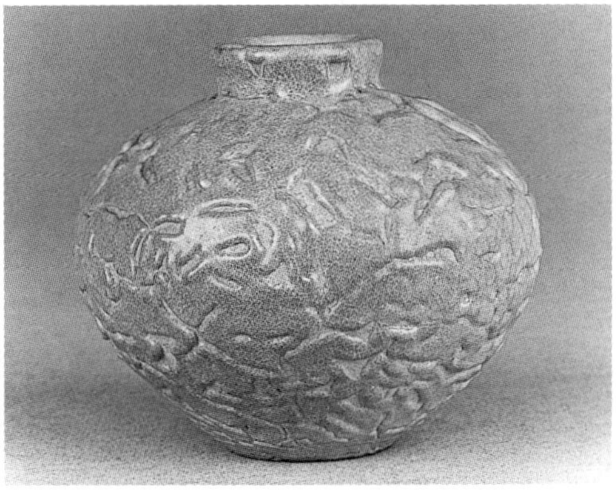

56. Small vase with crawled glaze. PDF Study Collection. Crawling occurs when a glaze does not adhere well to the body.

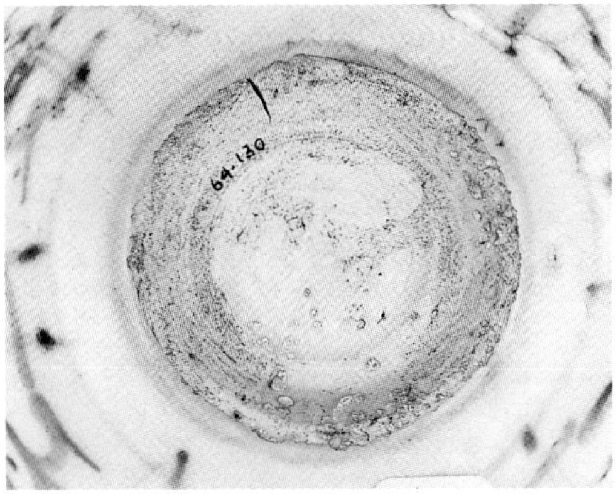

57. Base of a bowl with firing cracks. PDF Study Collection. Firing cracks are often caused by uneven drying or problems in the firing/cooling cycle.

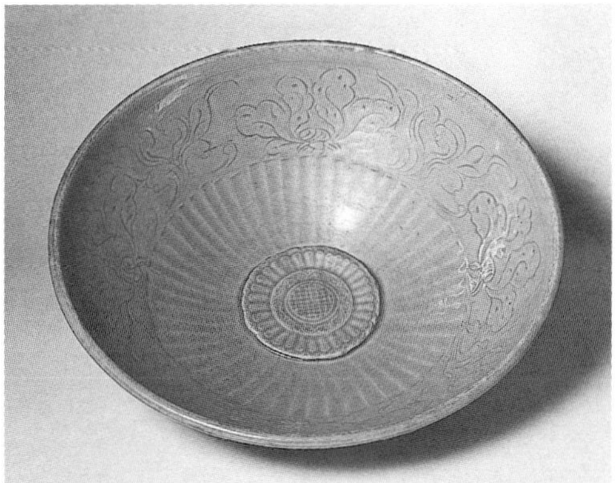

58. Longquan ware bowl with flower-shaped disk covering a hole cut in the base to allow for expansion and contraction during firing. PDF 277.

5. Drying and Firing

The process of changing wet clay into hard ceramic is called "ceramic change" and it occurs in the kiln during the firing. As we have seen in chapter one, water is added to crushed rock to make a clay body. This water lubricates the clay particles and allows them to slide past one another which enables us to shape a vessel from clay. In order to change this soft clay into a hard ceramic, this water must be removed and the clay particles must move closer together until they meet and eventually melt slightly so that they stick together permanently. The first stage in firing is actually done outside the kiln during the drying period after forming the pots. When a formed pot is left to dry, the added water or "water of plasticity" begins to evaporate into the surrounding atmosphere. The air in the workshop or drying area is usually warmed to speed up the drying process. This warm air contains less moisture than the pot so the water from the pot is drawn into the outside air until equilibrium is reached, that is, the amount of water in the air is the same as that in the pot. At this stage the pot feels warm to the touch. If the pot is still cold, there is still water left in it that needs to evaporate. When a pot is at the right stage of dryness, it is described as being "leather hard" because it feels like leather. The drying process is simple but it needs to be quite carefully controlled because evaporation caues water to move to the surface of the clay, which in turn causes the clay particles to move closer together. If a pot is dried too fast, therefore, it can crack. If it is not dry enough, when it is heated in the kiln too much steam can build up and the pot will burst. Often drying faults are not apparent until the pot is removed from the kiln so faults can be expensive.

As we have seen, clays with fine, small particles can hold more water than coarse, large particle clays. The more water present in a clay, the more there is to remove and thus the more it will shrink. A pot shrinks as it dries and this puts stress on the clay. So drying a pot too fast will cause it to shrink too fast and *vice versa*. Once a pot has dried, there is still some water left in the clay but not the added water. Some water is naturally present as "pore water", "adsorbed water", and "bound water" or chemically combined water. Pore water is locked inside the pores of dry clay[18] and is removed during the early stages of firing. Adsorbed water is present on the surfaces of the clay particles and therefore is also known as "interlayer water". Bound water is part of the chemical structure of the clay mineral and its removal in the later stages of firing is what helps to convert clay into ceramic. Neither pore water nor bound water is affected by drying.

In the firing process, a kiln is gradually heated up by increasing the temperature until the desired peak temperature is reached. As the kiln and thus the pot being fired are heating up, the firing

---

[18]Hamer, p. 339.

process proceeds through several stages. In the first stage, to about 120°C, the pore water is removed as steam. This stage is called the "water smoking" stage because the pore water has boiled away as "water smoke". This stage is tricky because if steam builds up too quickly, the pressure in the pot will rise and the pot will burst. Thinner pots will reach this stage quicker than thicker pots so this must be taken into consideration. At the end of this stage, the pots are "bone dry". In the next stage, from about 120°C to 350°C, any additional water , including adsorbed water is boiled off and vegetable or organic matter in the clay begins to decompose. This decomposition is accompanied by a slight expansion of the pot as objects expand with heat. From about 350° to 650°, ceramic change begins to occur. In this stage, the bound water is driven off as small amounts of steam and the clay mineral kaolinite begins to break down and is converted to another crystal called "metakaolin". The clay is no longer soft and has now become "ceramic", but a very weak ceramic at this stage because the clay particles themselves are now touching but not held together very strongly. It is in the next stage (650°C to 950°C) that any carbon is burned out and "sintering" begins which welds the clay particles together temporarily by softening the points of contact between them.[19] The burning out of carbon is necessary so that it does not form gas at a later stage and bloat the pot. From 950°C to about 1100°C, the process of "vitrification" begins. In this process, the clay crystal breaks down even further to form "spinel" and then free silica is released. This free silica melts and forms glass which glues the clay particles together to form a permanent bond. Vitrification means "turn to glass". The firing could be stopped in this stage to give an earthenware at 950°C or a stoneware body at 1100°C. However, if a truly strong ceramic is required (like porcelain), the firing must be taken a stage further so that a new crystal called mullite is formed. This needle-shaped crystal forms a lattice-like structure in the body which strengthens it. As mullite is formed, more silica is released and melts and thus more glass is formed in the body. This is why porcelain bodies ring when struck because they are so glassy. The firing in this stage can be taken up to about 1350°C.

After reaching peak temperature, the firing is often held at this stage to maintain the temperature for a period. This is called "soaking" and can be useful in allowing the glaze and body to mature without raising the temperature which might lead to overfiring. It also allows any bubbles in the glaze to smooth out. After soaking, the kiln is then cooled. During cooling, the ceramic body generally contracts at a regular rate and the glass formed in the body sets to form a solid.[20] Sometimes soaking is used during the cooling stage to allow crystals to grow in the glaze. In the firing process, therefore, water is removed from the clay and the pores start to close, bringing the clay particles closer together. The clay mineral begins to breakdown and change structure,

---

[19]Cuff, *Ceramic Technology for Potters and Sculptors*, p. 379.

[20]Hamer, *The Potter's Dictionary of Materials and Techniques*, p. 130.

eventually breaking down completely to form "ceramic" instead of clay. As the temperature increases, a new crystal is formed which adds strength to the ceramic. At this stage, the change is complete and the ceramic cannot be converted back into clay. During this change in the clay mineral, silica, a glass-former, is released and it melts which helps to fuse the ceramic together. This amount of activity does place a lot of stress on the pot as it is expanding and contracting throughout the firing process so firing needs to be carefully controlled. After the firing is finished, the pot is then left to cool. As it cools, it contracts so there is still a slight danger of cracking. Expansion and contraction are particularly problematic with very large vessels such as vases or bowls. An interesting method of dealing with this was used by potters at the Longquan kilns. In a large bowl, for example, a hole was cut into the base inside the foot ring. A glazed flower-shaped disk or medallion is then placed over the hole.(figure 58) The bowl was then placed in the kiln and fired as usual. As the body expanded and contracted and the glaze melted, the disk would float over the hole, falling into place only when the bowl cooled. The hole, therefore, gives the body room to expand and contract during the firing and the disk then seals the hole during the cooling. Holes were also cut into the bases of Dehua figures for a similar reason but these were not covered by a disk.

As we have seen in chapter two, in Chinese ceramics, the body and glaze of a vessel are usually fired at the same time. In most other ceramic traditions however, and with some Chinese ceramic types as well, the body needs to be fired first then glazed and fired a second time. This is usually the case when the body is high firing and the glaze is low firing, as with some lead-glazed wares. When the body is fired first, it is "biscuit fired". Biscuit firing is useful also for strengthening a very soft bodied ware. Whether a pot is biscuit fired first or glazed raw, it still must be handled carefully in the firing. When a glaze is fired, it goes through similar stages as the body but with more silica and glass-formers and the addition of fluxes to the mix, glazes are encouraged to melt and flow over the pot. When a pot is raw glazed, as with porcelain, the glaze and body mature together and a thick, integrated layer is formed at the body-glaze interface. Biscuit fired vessels are easier to handle when glazing and setting into the kiln but the "glaze fit" or integration between the body and glaze is not as good as with "green glazed" or raw glazed wares. This is why there appears to be no difference between the body and glaze in Chinese porcelain. As we have seen in some other high-fired wares, a layer of crystals can form at the glaze-body interface which can create interesting visual effects such as in Jun wares which have a layer of white anorthite crystals under the glaze which brightens it considerably.(see figure 20). In some types of ceramics, the glaze and body may be high-firing but the decoration may be low firing. This is the case whenever overglaze enamels are used to decorated a pot. These enamels are essentially low-firing lead based glazes that are applied to the glazed and fired vessel then fired again in a special low temperature kiln called a "muffle kiln". These enamel kilns are usually small and reach temperatures of about 800°C. Muffle kilns have a lining that acts as a giant saggar which protects the delicate enamels from the heat and dust in the kiln.

Any enamelled ceramic has been fired at least twice. Gold decoration requires a third firing at even lower temperatures.

As we have seen, a kiln is an enclosed chamber that contains air and has an atmosphere. This atmosphere changes when the kiln is fired up, depending on whether a lot of oxygen is let into the kiln or very little. When the fuel for the kiln is burned, the kiln master either closes all the flues or air holes in the kiln or keeps them open. If they are closed, the atmosphere becomes very smoky. If they are open, the atmosphere is nice and clean. As mentioned in chapter four, in a kiln, a smoky atmosphere is called a reducing atmosphere and a clean atmosphere is called an oxidizing atmosphere. In a smoky, reducing atmosphere, the air is starved of oxygen so it steals it from the pot. When this happens, any of the metal oxides in the pot, either for fluxes or colouring, are "reduced" to a lower state. Reduction firings tend to produce very cool and often greyish coloured wares. The opposite is true of oxidation firings which tend to produced warm and brightly coloured ceramics. The atmosphere not only affects the colour of ceramics but also their hardness, porosity and shrinkage. The atmosphere in a kiln is not constant, however. When the fuel is first lit, the atmosphere becomes smoky. When the firing finishes, the atmosphere clears and thus during cooling, it is oxidizing. So a pot can be fired in reduction but cooled in oxidation. This is one of the ways in which a potter can create cool and warm colours on one pot, as we have seen with the Longquan celadon wares with reddish biscuit decoration. (see figure 43 and chapter three for discussion) Oxidation and reduction are basic chemical processes that are constantly occurring in nature and the human body as well as ceramics. Other important factors in firing are the temperature of the firing and the duration. The temperature of the firing is what distinguishes the types of wares produced such as earthenware, stoneware or porcelain. It also determines to some degree the hardness of the ware. Certain changes in the body, such as the formation of anorthite crystals are dependant upon the duration of the firing as they can only occur if a high temperature was held for a long time. Thus, three main variables in firing must be considered when examining ceramics: temperature, duration and atmosphere. As we have seen, some celadons require short, quick firings whereas Jun wares require long, slow firings.

At Jingdezhen, a certain type of firing method was used that may or may not have been used elsewhere in China but should be mentioned as it is likely that other kilns using this method will be excavated in the future. This method is known as "water gas reduction" and is basically reduction that was water assisted. By dribbling water onto burning fuel, a very smoky steamy gas was produced (water gas) that burned very efficiently and helped to increase the heat in the kiln. This produced a strongly reducing atmosphere that was also efficient. The usual problem with reduction firings is that a reducing flame burns inefficiently so the temperature drops in the kiln. To raise the temperature, more fuel is required which is expensive. But with water gas reduction, this is not necessary as the temperature actually increases in the firing.

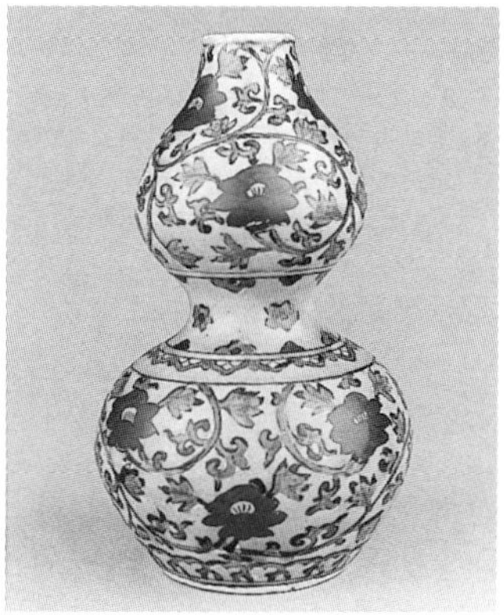

59. Double gourd shaped vase with overglaze enamel and underglaze blue decoration, Ming dynasty, Jiajing mark and period (1522-66). PDF 796.

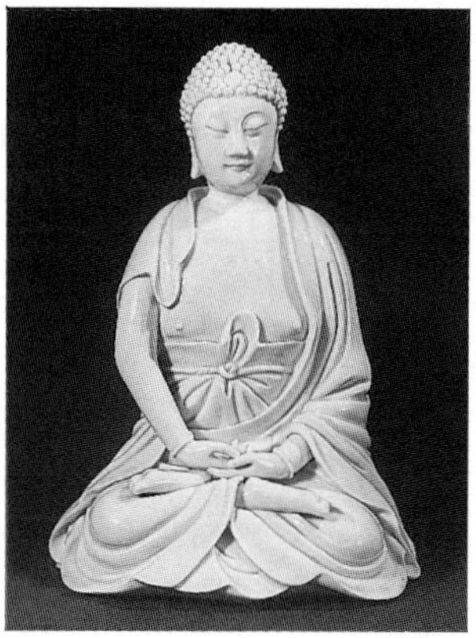

60. Dehua ware or *blanc de chine* figure of a Buddha, late Ming to early Qing dynasty, 17th century. PDF 488.

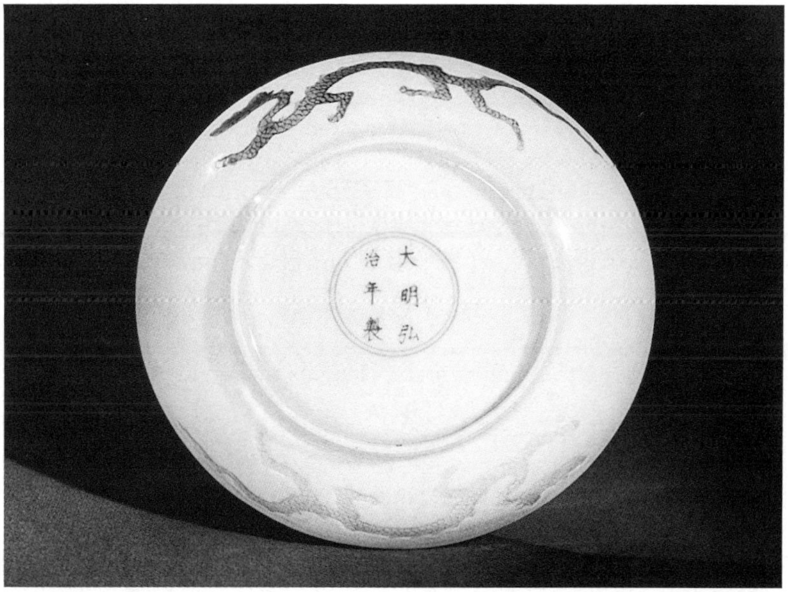

61. Dish with enamel-on-biscuit decoration, Ming dynasty, Hongzhi mark and period (1488-1505). PDF A724. The enamel colours used here are aubergine and yellow from manganese and iron.

List of Illustrations

30. Bowl with carved decoration, Longquan ware, Southern Song dynasty, 12th-13th centuries. D: 14 cm. PDF 263.
31. Bowl with incised decoration, *qingbai* ware, late Northern Song to early Southern Song dynasties, 12th century. D: 13.3 cm. PDF A459.
32. Dish with moulded decoration and mouth rim bound with copper alloy, Ding ware, Jin dynasty, 13th century. D: 21.5 cm. PDF 161.
33. Bowl with moulded decoration, Yaozhou ware, Northern Song dynasty, 12th century. D: 19.7 cm. PDF 245.
34. Dish with applied decoration, Jingdezhen porcelain, Yuan dynasty, mid-14th century. D: 15.7 cm. PDF A562.
35. Mould for a Ding ware bowl, dated by inscription to 1184. D: 21.9 cm. PDF 181.
36. Dish with slip decoration, Ding ware, Northern Song dynasty, 10th century. W: 13.4 cm. PDF 173.
37. Jar with slip painted decoration, Cizhou-type ware, Yuan dynasty (1279-1368). H: 24.8 cm. PDF 316.
38. Stoneware jar with slip decoration in the *fahua* style, Ming dynasty, 15th century. H: 35 cm. PDF 759.
39. Bowl with copper red glaze and slip decoration, Ming dynasty, Zhengde mark and period (1506-21). D: 16.9 cm. PDF A526.
40. Detail of a shard of *qingbai* ware with applied decoration, Yuan dynasty (1279-1368). PDF Study Collection.
41. Jar with underglaze blue and red painting and applied decoration, Jingdezhen porcelain, Yuan dynasty, early 14th century. H: 33 cm. PDF B661.
42. Porcelain bowl with reticulation and applied biscuit relief decoration, Ming dynasty, Wanli period (1573-1619). D: 9.5 cm. PDF A403.
43. Dish with sprig moulded biscuit decoration, Longquan ware, Yuan dynasty, 14th century. D: 43.1 cm. PDF 255.
44. Wine cup with overglaze enamel decoration in the *doucai* style, Ming dynasty, Chenghua mark and period (1465-87). D: 7.4 cm. PDF A745.
45. Dish with overglaze enamel decoration in the *wucai* style, Ming dynasty, Jiajing mark and period (1522-66). D: 17.2 cm. PDF A731.
46. Dish with overglaze enamel decoration in the *famille verte* style, Qing dynasty, Kangxi period, 1st half of the 18th century. D: 17.3 cm. PDF 805.
47. Detail of a dish with overglaze enamel decoration, early 18th century. D: 35 cm. PDF A821.
48. Flask with overglaze enamel decoration in the *famille rose* style, Qing dynasty, Yongzheng mark and period (1723-35). H: 29.3 cm. PDF 824.
49. Detail of figure 48 (PDF 824).
50. Detail of figure 48 (PDF 824).
51. Model of a Jingdezhen egg-shaped kiln, porcelain. H: 23.2 cm. W: 24.5 cm. Private Collection.
52. Line drawing of stepped saggars, after: *Kaogu* 1974, No. 6.
53. A selection of kiln furniture from the PDF Study Collection.
54. Base of a Ru ware brush washer showing spur marks, Northern Song Dynasty, early 12th century. PDF 76.
55. Shard of a dish with blistered glaze, modern. PDF Study Collection.
56. Small vase with crawled glaze, modern. PDF Study Collection. H: 7.7 cm. PDF Study Collection.
57. Base of a bowl with firing crack, Ming dynasty (1368-1644). PDF Study Collection.
58. Bowl with incised decoration and firing disk, Longquan ware, Yuan dynasty, 14th century. D: 41.9 cm. PDF 277.
59. Vase of double-gourd form with underglaze blue and overglaze enamel decoration, Ming dynasty, Jiajing mark and period (1522-66). H: 19.2 cm. PDF 796.
60. Seated Buddha, Dehua or *blanc de chine* ware, late Ming to early Qing dynasty, 17th century. H: 31.9 cm. PDF 488.
61. Dish with enamel-on-biscuit decoration, Ming dynasty, Hongzhi mark and period (1488-1505). D: 18 cm. PDF A724.

Selected Bibliography

Chang Foundation, *Imperial Hongwu and Yongle Porcelain Excavated at Jingdezhen*, Taipei, 1996.

Cuff, Y. H., *Ceramic Technology for Potters and Sculptors*, London/Philadelphia, 1995.

Fung Ping Shan Museum, *Greenwares from Zhejiang*, Hong Kong, 1994.

- *Ceramic Finds from Jingdezhen Kilns*, Hong Kong, 1993.

- *Dehua Wares*, Hong Kong, 1990.

Hamer & Hamer, *The Potter's Dictionary of Materials and Techniques*, third edition, London, 1991.

Hong Kong Urban Council, *Imperial Porcelain of the Yongle and Xuande Periods Excavated from the Site of the Ming Imperial Factory at Jingdezhen*, Hong Kong, 1989.

Kerr, R., *Chinese Ceramics: Porcelain of the Qing Dynasty 1644-1911*, London, 1986.

Kingery et. al., *Introduction to Ceramics*, 2nd edition, New York, 1976.

Kingery & Vandiver, *Ceramic Masterpieces*, New York, 1986.

Lam, *Tang Ceramics:Changsha Kilns*, Hong Kong, 1990.

Medley, M., *The Chinese Potter*, Oxford, 1989.

Mino Yutaka, *Freedom of Clay and Brush through Seven Centuries in Northern China: Tz'u-chou Type Wares, 960-1600 A.D.*, Indianapolis Museum of Art, 1980.

Mino and Tsiang, *Ice and Green Clouds: Traditions of Chinese Celadons*, Indianapolis Museum of Art, 1986.

Mowry, R., *Hare's Fur, Tortoiseshell, and Partridge Feathers: Chinese Brown and Black Glazed Ceramics, 400-1400*, Harvard University, 1996.

The Oriental Ceramic Society, *Iron in the Fire*, introduction by Nigel Wood, London, 1988.

Pellant, *Rocks and Minerals*, Eyewitness Handbooks, New York, 1992.

Rhodes, D., *Kilns: Design, Construction and Operation*, London, 1968.

Rice, P., *Pottery Analysis: a Sourcebook*, Chicago, 1987.

Scott, R., "18th Century Overglaze Enamels: the Influence of Technological Development on Painting Style" in *Style in the East Asian Tradition,* Colloquies on Art & Archaeology in Asia No. 14, PDF, 1987.

Scott, R., ed., *Chinese Copper Red Wares*, PDF Monograph Series No. 3, 1992.

- *The Porcelains of Jingdezhen*, Colloquies on Art & Archaeology in Asia No. 16, PDF, 1992.

Scott & Kerr, *Ceramic Evolution in the Middle Ming Period*, V&A/PDF, 1994.

Scott & Pierson, *Flawless Porcelains: Imperial Ceramics from the Reign of the Chenghua Emperor*, PDF, 1995.

Tichane, R., *Those Celadon Blues*, New York State Institute for Glaze Research, 1978.

- *Reds, Reds, Copper Reds*, New York Institute for Glaze Research, 1985.

- *Clay Bodies*, The New York Glaze Institute, 1990.

Tsui Museum of Art, *A Legacy of Chenghua*, Hong Kong, 1993.

Vainker, S., *Chinese Pottery and Porcelain*, British Museum, 1991.

Wang Qingzheng, *Underglaze Blue and Red*, Hong Kong, 1993.

- *The Discovery of the Ru Kiln*, Hong Kong, 1991.

Index